After Apollo

AFTER APOLLO

Cultural Legacies of the Race to the Moon

Edited by
J Bret Bennington and Rodney F. Hill

University of Florida Press
Gainesville

Published in the United States of America.

28 27 26 25 24 23 6 5 4 3 2 1

Library of Congress Cataloging-in-Publication Data
Names: Bennington, J. Bret, 1963– editor. | Hill, Rodney F., editor.
Title: After Apollo : cultural legacies of the race to the moon / edited by J Bret Bennington and Rodney F. Hill.
Description: Gainesville : University of Florida Press, [2023] | Includes bibliographical references and index. | Summary: "This book explores how NASA's space program impacted American society and culture during and after the race to the Moon, looking back at the 1969 Apollo 11 Moon landing from the perspective of the present day"— Provided by publisher.
Identifiers: LCCN 2022051316 (print) | LCCN 2022051317 (ebook) | ISBN 9781683403579 (hardback) | ISBN 9781683403692 (pdf)
Subjects: LCSH: United States. National Aeronautics and Space Administration—History. | Project Apollo (U.S.)—History. | Space flight to the moon—History. | Moon—Exploration. | BISAC: HISTORY / United States / 20th Century | TECHNOLOGY & ENGINEERING / Aeronautics & Astronautics
Classification: LCC TL799.M6 A37 2023 (print) | LCC TL799.M6 (ebook) | DDC 629.45/40973—dc23/eng/20221107
LC record available at https://lccn.loc.gov/2022051316
LC ebook record available at https://lccn.loc.gov/2022051317

University of Florida Press
2046 NE Waldo Road
Suite 2100
Gainesville, FL 32609
http://upress.ufl.edu

UF PRESS
UNIVERSITY OF FLORIDA

Contents

Illustrations

Figures

Table

Acknowledgments

The initial idea for this volume grew out of a series of special events commemorating the fiftieth anniversary of the Apollo 11 Moon landing, hosted in 2019 by Hofstra University's Cultural Center on Long Island—a mere stone's throw from the Cradle of Aviation Museum and not far from the former headquarters of Grumman Aerospace Corp., the birthplace of the Lunar Landing Module. The editors wish to thank a number of individuals at Hofstra University for their assistance and support: Melissa Connolly, Vice President, University Relations; Robert Brinkmann, former Vice Provost of Scholarship and Research and Dean of Graduate Studies; Athelene Collins, Executive Director of the Cultural Center; Carol Mallison, Cultural Center Events Manager; Jeannine Rinaldi, Senior Assistant to the Director of the Cultural Center; and all the other Cultural Center staff. The Apollo celebration was made possible by the Joseph G. Astman Distinguished Scholar Fund for the Hofstra Cultural Center and by the Office of the Provost and Senior Vice President of Academic Affairs. We also wish to acknowledge the new President of Hofstra University, Dr. Susan Poser, whose commitment to research and scholarship bodes well for similar endeavors in the future.

Of course, we are most appreciative of the level of support that we have received from the University Press of Florida, particularly from Senior Editor Sian Hunter, whose patience with us as we struggled to make progress on this volume while managing the challenges of running our respective academic departments during a pandemic is gratefully acknowledged.

* * *

J Bret Bennington would also like to thank his coeditor, Rodney F. Hill, for being such a terrific collaborator on this project. He did so much of the heavy lifting, communicating with authors and editing manuscripts. I would also like to thank my colleagues in the Department of Geology, Environment, and Sustainability at Hofstra and in other departments across campus for their enthusiasm for working across disciplines and for always indulging me when I ask for help with projects. My horizons have been expanded in many directions because my friends and colleagues at Hofstra have invited me along on any number of intellectual adventures. In my department, I thank Christa Farmer, Antonios Marsellos, Jase Bernhardt, and Deirdre Kujawski for their support and camaraderie. Stephen Lawrence in Physics and Astronomy has been my go-to guide for all matters extraterrestrial, while Sabrina Sobel in Chemistry is always willing to mix it up to see what precipitates. Thanks also to Stavros Valenti for moral support, to Russell Burke for indulging the herpetologist in me, and to Lauren Burignat-Kozol for introducing me to the music of the spheres. Finally, my most heartfelt thanks go out to my wonderful wife, Gail Bennington, and our girls, Maureen, Fiona, Danaan, and Kylie, for putting up with my working late nights at home and being distracted and preoccupied so much of the time.

* * *

Rodney F. Hill would like to thank his coeditor, Dr. J Bret Bennington, for making such tremendous contributions to this book, which far exceed what Prof. Bennington's modest remarks above might suggest. Further gratitude goes to my colleagues in Hofstra's Department of Radio, Television, Film, especially Phil Katzman, Bill Jennings, Christine Noschese, Paula Uruburu, George Nicholas, Aashish Kumar, Carlo Gennarelli, Randy Hillebrand, Aaron Braun, Russ Harbaugh, Karl Galvan, Mary Otarola, and Nina Rivera. I am also grateful to Dean Mark Lukasiewicz of the Lawrence Herbert School of Communication and everyone in the Dean's Office: Mario Murillo, Adria Marlowe, Michele Roberts, Robyn Crosby, Susan Mulligan, and David Henne. Nothing in the Herbert School is possible without the support of our ace facilities staff: Tim Fehmel, Cameron Keough, Victor Lazo, Patricia Szenher, and Fred O'Neill. I am also thankful for the outstanding students, past and present, in Hofstra's film and television programs, who inspire me constantly, especially: Connie Tais, Muhammad Muzammal, Spencer Fowler, Margaret Hart, Charlie Norton, Chika Okuyama, and Santiago Cardona. My work on this book is dedicated to J. P. Telotte and in memory of James M. Welsh. On a personal note, I am forever indebted to dear friends and family: Abraham Albert and Michael Cooper, Todd Jones, John C. Tibbetts and Mary Lou Pagano, Wendy Lidell, Lance Schwulst and Nathan Tsoi, Joe Yranski, George and Kachina Spyros, Rick and Marian Clancy, Judy and Johannes Scharf, Patti Hill, Dylan Adams, and Elaine Hill.

Introduction

J BRET BENNINGTON

On July 20, 1969, an ungainly looking craft constructed out of aluminum, Mylar, and gold foil, the first vehicle ever built to operate solely in the vacuum of space, touched down on the surface of the Moon with two American astronauts inside. As over 500 million people watched live on television around the world, first Neil Armstrong and then Buzz Aldrin emerged from the spacecraft and became the first humans to set foot on another world. The engraved, stainless steel plate fastened to one leg of the lunar lander proclaimed, "WE CAME IN PEACE FOR ALL MANKIND." Although an American technological achievement, the Apollo 11 Moon landing was heralded as a milestone in the progress of the human species. The essays collected in this volume, written to commemorate and expound on the fiftieth anniversary of the first Moon landing, explore this achievement and its impacts on people from a variety of unique, multidisciplinary perspectives.

To break free of the Earth's gravitational attraction and launch Apollo 11 on its journey to the Moon, the powerful Saturn V rocket had to accelerate to a speed of more than 24,000 mph. This tremendous acceleration is an apt metaphor to describe how the world changed in the aftermath of World War II. Indeed, economists and sustainability scientists refer to the postwar period as the "Great Acceleration," when everything from global population and resource consumption to computing power and the number of fast-food restaurants began climbing along an exponential curve. Never before had civilization experienced such a rapid rate of change. Entering the third decade of the twenty-first century, we are still riding this rocket, collectively straining against the g-forces of globalization, automation, connectivity, and climate change.

We are now more than fifty years past the first Moon landing, but humans have not yet returned to the Moon—the last astronauts to wander the lunar surface departed with Apollo 17 on December 14, 1972. In a way, Apollo was a victim of its own success, so thoroughly besting the Soviets that it ended the Space Race. Even the near disaster of Apollo 13 only reinforced the impression that American know-how could solve any problem. The public quickly lost interest as Moon landings became seemingly routine. Without strong public interest, the political equation changed, and the expense of Apollo, which never sat well with the American public, became unsupportable in a nation struggling with poverty and the Vietnam War. President Nixon, facing increasing political pressure, canceled the Apollo program, and NASA shifted its attention and resources to robotic explorations of the solar system and manned missions to low Earth orbit designed to tackle earthbound problems.

However, the changes set in motion by the Space Race and the Moon landings have not stopped. For those of us who remember watching Neil Armstrong step down onto the lunar surface, probably on a small black-and-white TV screen, that moment seemed like the beginning of a new era, when science fiction became fact and all bets on the future were suddenly off. The world was already on its ballistic trajectory, but few probably realized how much life was changing and how quickly. The Moon landing brought that reality home, beamed from the lunar surface directly into our living rooms.

In the intervening decades, the Apollo 11 Moon landing has been the subject of voluminous studies in various fields of science, engineering, history, political science, journalism, and cultural studies—as well as the inspiration for speculative writings in science fiction, comic books, film and television scripts, and other media. This collection represents a rather unique approach to this endlessly fascinating topic in that it brings together a group of scholars from disparate fields—many represented above—for an interdisciplinary reflection upon Apollo 11 and the many major impacts that it has had on people. From the astronauts in their spacesuits walking on the Moon to the immigrant engineers working to get them there, from the women who embraced astronaut culture yet struggled to gain acceptance to the space program, to the civil-rights activists who rejected spaceflight as a profligate waste of resources—what unites these essays is a focus on the many ways that different kinds of people have experienced and been impacted by the novel phenomenon of spaceflight leading up to Apollo 11 and beyond.

The sheer improbability of the Moon landing is explored in Matthew H.

Hersch's essay, "'We Ran as if to Meet the Moon': The Inspired Lunacy of Apollo 11," which provides an insightful overview of the entire Apollo project and its origins. John F. Kennedy's pledge in 1961 "that this nation should commit itself to achieving the goal, before this decade is out, of landing a man on the Moon and returning him safely to the Earth" was an audacious gamble, given that the technologies needed were embryonic at the time. Nevertheless, as Hersch explains, the United States out-collectivized the Soviets to muster the resources and expertise needed to accelerate the development of human spaceflight (a Cold War application of the government-industry partnerships developed during the heat of World War II). Apollo 11 was the culmination of "project astronaut" (Hersch), NASA's program to craft a new breed of explorer out of military fighter pilots. By the end of Apollo, the narrative of outer space would shift from conquerors to colonizers, as engineers and scientists increasingly joined the astronaut corps (Hersch; Spiller, this volume).

Although the scientific legacy of Apollo is often sought in aeronautical engineering or planetary geology (e.g., Crotts), the race to the Moon touched all aspects of science and technology. Some of the chemistry and material science innovations developed for the Moon landing are explored by Sabrina G. Sobel in "The Hidden Chemistry behind the Apollo 11 Mission." For many, the term "spin-offs" as applied to the space program may evoke Tang and personal computers, but the range of technologies developed to land on the Moon and their reach into our lives is vastly broader. Sobel provides some examples in her discussion of how chemistry was critical to the invention of giant liquid-fueled rockets (both the fuel itself and the technology of handling it), the new synthetic materials needed to make spacesuits, and personal metabolic waste management in the confines of a spacecraft. Out of these novel Apollo technologies developed diverse applications that have become routine aspects of modern life, such as medical MRI, thermally protective clothing, and Mylar party balloons.

Solving the technological problems posed by spaceflight was only one of the primary obstacles to getting to the Moon. Convincing the American public that this endeavor was important to the nation and a justifiable expense of taxpayer money was perhaps the greater hurdle. While the existential panic sparked by the Soviet Union's successful launch of Sputnik into Earth's orbit in 1957 certainly made it easier to justify public funding for STEM education, scientific research, and the development of space technologies, it was German propaganda that was already at work convincing

Americans that the path to our destiny led through interplanetary space. As Chris Robinson documents in "Selling Space Travel: The Disney Version," former Nazi V-2 rocket scientists, spirited from Germany at the close of World War II, brought to America their dreams of exploring space along with their technological expertise. Chief among these immigrants was Dr. Wernher von Braun, who would go on to design the liquid-fueled rockets that carried men to the Moon. Von Braun evangelized space travel in a series of articles for *Collier's* magazine that eventually led him to collaborate with Walt Disney on three episodes of the popular television show *Disneyland* in the mid-1950s. These shows attracted large audiences and featured von Braun and other scientists selling the idea that spaceflight was a desirable goal within reach. Von Braun's vision for human spaceflight was also reflected in the Tomorrowland section of the new Disneyland theme park, including a life-size model of a TWA-branded Moonliner designed with input from von Braun. The role of von Braun, *Collier's*, and Disney in making spaceflight seem real to the American public has been discussed elsewhere (McCurdy), but Robinson also suggests that, in addition to priming the American public for space travel, Disney's "Man in Space" episode was used by President Eisenhower to help convince skeptical generals of the need for a government program to develop the capability to launch satellites. When the Soviet Union launched Sputnik into orbit, the American satellite effort was well underway, and Explorer 1 was launched only three months later. Thanks to von Braun and Disney, many American taxpayers were already space enthusiasts by the time Sputnik was circling the Earth overhead.

Nevertheless, it would be a mistake to assume that the average American was fully in favor of the enormous expenditures of federal money fueling the American space program. As Patricia Rossi documents in her contribution, "The Launch of Apollo 11: A Necessary Mission in the Space Race vs. Superfluous Government Spending," only about one in three Americans supported the expense of the Apollo effort. The skepticism over spending billions of dollars to put men on the Moon was strongest among African Americans. Civil rights leaders such as Martin Luther King Jr. and the Reverend Ralph Abernathy were vocal opponents of the Apollo program. Although respectful of the intentions and achievements of the NASA astronauts and scientists, most African Americans were incensed that the United States could direct so much money to space exploration when the needs of Black communities were so great and so many Americans lived in poverty and squalor. As the poem by Gil Scott-Heron vividly expressed it, "A rat done bit my sister Nell with Whitey on the Moon . . . Her face and arms began to swell . . . Whitey on the Moon . . . I can't pay no doctor bill, but Whitey on the Moon. Ten year from now I'll be payin' still . . . Whitey on the Moon" (Madrigal).

Rossi shows how the widespread opposition to extravagant space expenditures had a significant political impact and likely was part of the motivation behind President Johnson's War on Poverty and associated Great Society legislation. By the time of the Apollo Moon landing, political momentum was building to cut NASA's budget and refocus government funding toward improving conditions at home. Politicians who started out supporting the space program became critics—even JFK's brother Senator Ted Kennedy came out in favor of redirecting NASA dollars in a speech just months before Apollo 11 launched. Given these changing political winds, it is not surprising that the Nixon administration chose to cancel the Apollo program and redirect a defunded NASA toward missions aimed at solving environmental problems on Earth.

Although one can argue that the Space Race both helped and hurt the circumstances of minorities in the United States, NASA and the space program were a boon to immigrants, who made significant contributions to the science and engineering workforce behind the Moon landings. In her essay, "Scientists Without Borders: Immigrants in NASA and the Apollo Program," Rosanna Perotti argues that there are close connections to be found in the risks and hardships of immigration and space exploration. She documents how the space exploration community became a model of diversity and multiculturalism, thanks to the high proportion of scientists and engineers who come to the United States to pursue advanced degrees and then elect to remain and work in this country. Perotti provides compelling anecdotes of the important work done by immigrants and first-generation Americans in the conquest of outer space. At first, the immigrant workforce at NASA was composed primarily of Europeans seeking refuge from the political and social upheavals of postwar Europe, including German scientists such as Wernher von Braun who were brought by the Army to the United States in Operation Paperclip to continue their rocketry work. However, after changes to national immigration policy opened up the flow of immigrants from outside of Europe, the foreign-born technical workforce began to diversify. One of the scientific heroes of the Apollo program profiled by Perotti is Farouk El-Baz, an Egyptian immigrant who trained the Apollo astronauts in making geological observations from lunar orbit. Regarded at first with a certain level of skepticism, El-Baz went on to become a beloved scientific mentor to the men who went to the Moon.

As the space program matured in the years beyond Apollo, immigrants and their children continued to make important contributions to satellite missions to Earth and robotic exploration of the planets. Immigrants also made their way into the astronaut corps to become Space Shuttle and International Space Station astronauts. Even the current push to privatize space exploration has its share of notable immigrant talent, including SpaceX's Elon Musk, whose country of origin is South Africa (Spiller, this volume). As NASA prepares to return to the Moon for Project Artemis, immigrant Americans are in the running to be part of the second wave of lunar exploration.

Just as African Americans and immigrants were initially excluded from becoming astronauts (all of the Mercury, Gemini, and Apollo astronauts were white, native-born American males), so too were women. Julie Wosk in her contribution, "Picturing Women in the Space Age: The Impact of the Lunar Landing on Films, Television, and Fashion," tells the story of how women were influenced by and incorporated into the culture of space exploration through the lens of fashion and depictions of "space women" in television and film. The popular interest in space exploration created in the 1950s and 1960s was appropriated by fashion designers who dressed women in spacesuits and outer-space-themed garments long before any women were allowed to don a flight suit on an actual mission. Wosk documents how actual women in space were preceded by "space women" in advertising, fashion magazines, TV, and films. Even in the 1950s, when the model for women in America was usually the dutiful housewife (Jane Jetson provides an apt space-age example), women were depicted in TV serials such as *Space Patrol* as capable space pilots and adventurers. In the 1960s the ever-visionary *Star Trek* presented Lt. Uhura to Americans as an African American communications officer serving on a United Federation of Planets starship. Dr. Mae Jemison, the Shuttle astronaut who was the first African American woman in space, credits the character of Uhura as an important influence on her interest in becoming an astronaut. In the twenty-first century, women are increasingly seen as commanders and heroes in space, both in fiction and in fact. NASA is now preparing to send astronauts back to the Moon in the Artemis Program (named for the

Greek goddess who was Apollo's twin sister), and they plan to land the first American woman on the Moon as part of the crew of Artemis III—the first *peopled* lunar landing since 1972.

To date, twelve men have walked on the Moon. The direct, first-hand experience of being on the surface of another planet now resides in the memories of the four Apollo astronauts who are still with us. Allison Whitney explores how Hollywood has attempted to recreate their experiences for audiences through the medium of IMAX films. In her essay "'His Own Personal Adventure': Lunar Exploration and the IMAX Experience in *Magnificent Desolation* and *First Man,*" Whitney deconstructs the cinematography of these two films to better understand how they were used to convey realistically and emotionally what is perhaps the most unique and exclusive of human experiences. She also explores how photography was itself an integral part of the way that the astronauts experienced their journey to the Moon and how the IMAX film format has become the preferred cinematic vehicle for space exploration documentaries. Indeed, NASA has entered into a collaboration with an IMAX subsidiary, IMAX Space Ltd., to "design cameras for use in space, train astronauts as IMAX cinematographers, and create films that inspire space science." As with Disney in the 1950s, NASA is again using popular entertainment to promote a vision of the future centered on human spaceflight. At the end of *Magnificent Desolation*, viewers are shown an imagined future where a young girl who appeared in an earlier sequence envisions herself as an adult living on a Moon base, planting the seeds of the next generation of space women to grow up with dreams of human space exploration.

Artemis is being justified, in part, as a stepping stone to the human exploration of Mars. There are once again elements of a Space Race, only now our competition is coming from China as much as from Russia. Unlike in the 1960s, the major driver of the current advances in human spaceflight is coming from the push to commercialize space. The leading names in commercial human spaceflight, *SpaceX*, *Blue Origin*, and *Virgin Galactic*, are betting on space tourism to provide the revenue needed to pay for the technological innovations that will make possible the human colonization of interplanetary space. James A. Spiller examines the post-Apollo shift from a frontier narrative to tourism as the justification for an American presence in space in his essay, "America's 'Space Frontier' in an Era of Space Tourism." Spiller provides a historical analysis of the American mythologies employed by proponents of an expanding human presence in space and the exploration and colonization of other planets. Whether equating the Apollo astronauts to explorers of new territories or Shuttle astronauts to pioneers going to work in the frontier of space, Americans have always viewed space exploration as a distinctly American prerogative and expression of our national character. Spiller questions whether the mystique of spaceflight can survive the commercialization of space. Private companies such as SpaceX will most likely continue to require significant government subsidies to develop the technologies needed to establish a permanent human presence on the Moon and Mars. If the new model astronaut is a millionaire tourist paying for a sightseeing tour, will the public still be willing to contribute tax dollars to this effort?

It remains to be seen whether there are opportunities to profit from space beyond Earth's orbit. If fortunes can be made colonizing the Moon and Mars, then it seems highly likely that they will eventually be colonized. However, if the payoff of exploring the solar system ends up being exclusively scientific, then commercial space travel may be inherently self-limiting. Or perhaps there is a third possibility that combines the two. If wealthy visionaries like Elon Musk, Jeff Bezos, and Richard Branson are willing to subsidize the myth of the space frontier with dollars made on Earth and from space tourism, then it seems likely that there will always be people ready to lend their expertise, effort, and even their lives to space exploration—ready to take that "one giant leap" to expand the human presence beyond planet Earth. As the papers collected in the volume demonstrate, the desire to go to space has always been a multifaceted expression of the human experience.

Works Cited

Crotts, A. *The New Moon: Water, Exploration, and Future Habitation*. Cambridge UP, 2014.

Hersch, Matthew H. *Inventing the American Astronaut*, Palgrave Studies in the History of Science and Technology. Palgrave Macmillan, 2012.

Madrigal, Alexis C. "Gil Scott-Heron's Poem, 'Whitey on the Moon.'" *The Atlantic*, 28 May 2011.

McCurdy, Howard E. *Space and the American Imagination*. Smithsonian Institution Press, 1997.

1

"We Ran as if to Meet the Moon"

The Inspired Lunacy of Apollo 11

MATTHEW H. HERSCH

The most astonishing wonder seen by the first humans who looked through two glass lenses placed at opposite ends of a paper tube was not the rings of Saturn, the satellites of Jupiter, or the phases of Venus (each of which had their own startling scientific implications), but Earth's humble, constant companion, the Moon. Taught to believe that the Moon was a flawless, divine orb, anybody with a telescope in the first decade of the seventeenth century suddenly knew the truth: the Moon was a world, perhaps like ours, with mountains, valleys, and rolling plains inviting visitation from Earth. Humans had imagined traveling to the heavens for millennia, but the telescope gave them a destination worth visiting—not a distant point of light, but a place to live. Certainly, the journey would be arduous, and all who attempted it could not be expected to survive. Yet even before thinkers possessed a craft capable of making the journey, they started to imagine who should make it.

Pioneering German astronomical theorist Johannes Kepler, writing in his posthumously published science fiction story, *The Dream*, scoured Europe for appropriate crew members for such a journey. "We admit to this company nobody who is lethargic, fat, or tender. On the contrary, we choose those who spend their time in the constant practice of horsemanship or often sail to the Indies, inured to subsisting on hardtack, garlic, dried fish and unappetizing victuals. . . . No men from Germany are acceptable; we do not spurn the firm bodies of the Spaniards" (Kepler and Rosen 15). Kepler also praised middle-aged women, whom he regarded as inured to all manner of suffering, perhaps a tribute to Kepler's own mother, an accused witch with whom he was close throughout his life. Unable to conjure, in his tale, a machine capable of making the trip, Kepler sought help from his mother's magic, but the problem of propulsion to the Moon would not be solved by witchcraft.

Sir Isaac Newton, fifty years later, put Kepler's musings on more solid scientific footing by describing how small, simple objects might be hurled into space, though piloted vehicles remained a distant dream. Rather, Newton hypothesized about cannonballs fired off of high mountains so fast they might fall perpetually around the curvature of the Earth and orbit it, or escape the gravitational pull of Earth altogether and reach another astronomical body (Newton et al.). Getting to the Moon—achieving speeds measured in the tens of thousands of miles per hour and crossing a distance that even the ancient Greeks knew to be of hundreds of thousands of miles—would not be easy, and indeed, no technology of the seventeenth century could send a projectile even a few miles into the sky, let alone blast a cannonball into orbit. While Kepler had imagined summoning the aid of magic to complete the journey, Newton abandoned the problem and busied himself with alchemy and biblical study, while other thinkers reached out for new forms of propulsion (Westfall).

The closest thing to a spaceship that the seventeenth century could summon could be found in China, where a century-old and likely apocryphal account chronicled the adventures of an astronomer of the Imperial court. Wan Hu, it is said, fastened fireworks to his chair and ordered his assistant to light their fuses. Once the smoke had cleared, Wan Hu had disappeared, although it is not clear to where, or in how many pieces. What can we learn from this story, or, indeed, from the countless other tales of real people whose conviction that human beings could fly were so great that they risked and ultimately surrendered their lives to this one peculiar dream? Exploration of the heavens has, throughout history, come at great cost, from the ill-fated mythological flight of Icarus to unlucky nineteenth-century glider pioneer Otto Lilienthal, whose fragile body was no match for the wind. Attempting to reach the heavens, humans have been told in religion and myth throughout time, is not allowed by the gods, and yet they persisted.

When I consider the story of Apollo 11's voyage to the Moon more than fifty years ago, I am struck, not merely by the difficulty of the journey, but by the gargantuan, and almost absurd, confidence, possessed by humans for centuries, that the Moon was a place that humans could and would travel to, and that dangers to which they would expose themselves while trying were well worth the risk. I am drawn less to the stories of how the

Apollo 11 astronauts feared that they might fail in their journey than to their surprising conviction that they would succeed. To modern audiences, their faith seems foolhardy, but mid-twentieth-century Americans lived in an age of miracles, in which almost anything seemed possible.

While nations around the world have contributed to the effort to explore space (and achieved noteworthy successes), it was only the peculiar mobilization of talent and technology by the United States in the middle decades of the twentieth century that allowed humans to walk on the Moon. While Cold War fears of the Soviet Union motivated the huge expenditures required for Apollo 11 to succeed (McDougall), the necessary conditions for that triumph had been established much earlier. At a seemingly hopeless moment early in the Second World War, the US forged a new kind of national innovation policy that made the federal government a direct consumer of science and technology produced by universities and private companies. Using capital provided by taxpayers, entities like the Office of Scientific Research and Development created the arsenal of democracy, defeated fascism in Europe and Asia, and then transformed the Moon from a distant object of wonder into a place that people could actually visit (Baxter).

When critics on the American Right condemned Project Apollo as a kind of creeping socialism (McDougall), they were more right than wrong, a charge made even more galling by the fact that it worked. Americans, as it turned out, were better collectivists than the Soviets were. The US government was not an incompetent and wasteful bureaucracy, but a storehouse of expert knowledge and civic-mindedness (Webb; Lambright; Johnson). Private corporations contributed their own funds to advance technologies they believed valuable and accepted modest profits for their efforts. Though it gave the appearance of a populist paradise, the Soviet Union was, by contrast, a corrupt and ineffective oligarchy, divorced from truth and devoted to enriching its leaders. It interfered in the work of its bickering rocket design bureaus and forced them to compete ruinously among themselves, collapsing into blame and recriminations when its efforts failed (Siddiqi).

As a program requiring selfless commitment to a common goal and the overcoming of dozens of scientific and technical challenges, Project Apollo, to modern audiences, should not have succeeded. That it did owed much to the fact that landing on the Moon was itself a sixty-year journey that began with heavier-than-air flight and continued through a traumatic global conflict that reconfigured popular notions of the possible. Rocketeers of the early twentieth century were an international fraternity of theoreticians

Figure 1.1. Dr. Robert Goddard and his team pose with one of Goddard's liquid-fuel rockets in 1932 (NASA Goddard Space Flight Center photo).

and experientialists, though it was American college professor Robert Goddard who, in 1926, achieved the first flight of a liquid-fueled rocket: the only propulsion technology with the power to propel humans to the Moon (Lehman) (see fig 1.1). This technology found application elsewhere, notably in Adolf Hitler's Germany (Neufeld), but while the later contributions by German engineers to the American space program were significant, the seeds of Apollo lay in American innovations before and during World War II.

From ground zero of the global aircraft industry on Long Island, companies like Vought, Republic, and Grumman—entities that would someday build rockets and spaceships—built propeller planes and jets, while elsewhere, test pilots chipped away at the problems of high-speed, high-altitude flight (Bilstein, *Flight in America*). The people who built and flew these airplanes through World War II and the years that followed went on to build and fly the spacecraft of Project Apollo, while elsewhere, the war against fascism provided funds that turned a number of notional technologies into embryonic ones like spacesuits, hypothesized by American science fiction author Garrett Serviss in 1898 but first constructed to protect the crews of high-altitude wartime bombers (Serviss; Mallan). Composed of accordion-like rubber joints and Plexiglas helmets, early models by tire maker B. F. Goodrich expanded when inflated and proved almost impossible to move around in. Indeed, scarcely any of the technologies that sent Americans to the Moon in 1969—rockets, radar, computers—did not exist in some form twenty years earlier, though they were, in many cases, cumbersome and impractical rough drafts rather than working devices. Though not yet feasible technologies, they provided enough inspiration to enable futurists to imagine that humans would use them to travel to the Moon within their own lifetimes.

In 1949, *LIFE* magazine imagined a lunar voyage undertaken in 1974, only five years after Apollo 11's flight. Rather than imagining fanciful new technologies, *LIFE* artist Noel Sickles assembled real American weapons and industrial design of 1949 to craft a convincing visual representation of what spaceflight would look like twenty-five years in the future. The aesthetic vision of the *LIFE* images was of a peculiar form of American military style, found everywhere from Pentagon offices to battleships in 1949: gray office chairs with green vinyl padding, steel sleeping berths, war-surplus radar displays, and a design aesthetic reminiscent of austere wartime life, not the Art Deco fantasies of prewar science fiction ("Rocket to the Moon: Man May Travel to Earth's Satellite within 25 Years"). If anything, Sickles's vision was conservative: none of the machines or devices he depicted were fanciful, and the fashions and even the hairstyles of the men seem similarly frozen in time to a period just following World War II.

If there was ever a population comfortable with the idea of space travel, it was postwar America. In children's books and popular films of the era, an entire generation was groomed to be Moon voyagers long before Apollo 11 flew (M. B. Freeman). With the threat of Soviet expansion and Soviet technology looming on the horizon, flight into space became a matter of national security, not because space appeared to be a place where wars might be fought, but because mastery of the technology to get there appeared inextricably entwined with the technologies needed to defend oneself on Earth, and would, even more importantly, signal to nations that had not yet chosen to favor the West or East which system of government would control the future.

The Soviet Union's early accomplishments in space—launching the first artificial Earth satellite, the first probe to photograph the lunar far side, and the first human into space—presented the United States with what appeared to be a desperate but nonviolent battle for the hearts and minds of the world. By mid-1961, eight years before Apollo 11, Americans had launched only a single human astronaut, in his Project Mercury spacecraft, into space, and only for fifteen minutes, neither orbiting the Earth nor venturing more than about 100 miles from its surface (Swenson Jr. et al.). Many people in this period had thought about flights to the Moon, and some, especially within the Army and Air Force, had even begun to sketch out how such a flight might occur, but it was a goal for another time, offering too many scientific unknowns and unsolved technological challenges.

This dissonance between desire and available technology made President John F. Kennedy's Urgent National Needs address of May 25, 1961, all the more jarring. Determined to offer Americans a space exploration goal so audacious that any temporary Soviet superiority in space would confer no advantage, the National Aeronautics and Space Administration and the White House fixed upon piloted lunar landing and return, effectively promising that the nation's space program would swim the Long Island Sound after mastering the doggy paddle. Whether Kennedy ever expected to achieve the goal of landing a man (women would not be considered for the program) on the Moon and returning him safely to Earth by 1970 is unclear; evidence suggests that he regretted the suggestion almost immediately, and political talk for the next two years centered upon whether Congress would appropriate the necessary funds, or whether the US might need to collaborate with the Soviet Union on such a flight (McDougall). Though he had no personal interest in space exploration, Kennedy, however, understood the space program's geopolitical importance and warned his staff of the dangers to American prestige if the nation were to come up last in the Moon Race. Even if Kennedy's challenge had been a political ruse—a Potemkin space project intended only to discourage competition and boost his popularity—nobody let NASA in on the joke. Having been chartered by Congress in 1958 as a peaceful scientific organization with a broad purview and no specific goals, it found, in Kennedy's challenge, a new mission and a new relevance, one that would prove both a blessing and a curse (Launius).

In the decade that followed, NASA pursued a lunar landing with maniacal intensity, all the more impressive given the obstacles in its path. In the year of NASA's creation, the vehicle that would carry men to the Moon—the liquid-fueled rocket—seldom worked more than one time in a dozen. To build a launch vehicle capable of carrying spacecraft larger and heavier than anything previously sent into space (and not merely into Earth's orbit, but to the Moon and back) required a rocket of a size and reliability that existed only in engineers' imaginations. Rigorous testing of subsystems and expensive failures slowly improved reliability, while engineers under former Nazi rocketeer Wernher von Braun at NASA's Marshall Space Flight Center cobbled together the tanks and engines of its smaller Redstone and Jupiter rockets to create a cluster rocket that could serve as the first stage of a launch vehicle that, with a little luck, might just barely get half of a Moon ship less than a tenth of the way to the Moon.

Beginning with the humble Saturn I, NASA and its contractors devised a series of ever larger rockets dwarfing even this behemoth in size and mass (Bilstein, *Stages to Saturn: A Technological History of the Apollo/Saturn Launch Vehicles*). Sending a spacecraft into Earth's orbit was a relatively simple matter, and to return from orbit, a spacecraft needed only a breaking rocket to slow it just enough to fall into the thicker layers of the atmosphere where atmospheric drag would do the rest of the work. A round trip to the Moon would require more feats of engineering: a spacecraft that could not only fly into space, but fly to the Moon, slow down, land, and then lift off the Moon's surface and return to Earth, taking with it all of the fuel required for the journey.

The round-trip journey would take at least a week and require a crew of three if the crew members were to take turns sleeping (French and Burgess 215). Even with the smallest spacecraft capable of making the journey and returning to Earth (a conical Project Apollo capsule with a three-person crew), the craft that landed on the Moon would be a monster. Astronauts, lying on their backs, would descend to a lunar surface they could not see, and then climb down a long ladder to the lunar surface. Any damage to the craft would likely doom the astronauts to remain on the Moon. This "Direct Ascent" method offered some measure of simplicity but required a rocket, called Nova, so large that it would stretch Kennedy's timetable and potentially put a lunar landing out of reach.

It was NASA engineer John Houbolt who, bucking the prevailing wisdom of von Braun and others at NASA, pushed a new method of landing: Lunar Orbit Rendezvous, in which NASA would dispatch to the Moon both a command ship and a smaller lander, which would descend to the lunar surface and return with a portion of the crew, completing the landing with a much smaller rocket than that required for Direct Ascent (Brooks et al. 68). Without the need to carry as much fuel, Apollo's twin craft would weigh less than a single spacecraft intended to fulfill all parts of the flight.

The mission would still require only a single launch, but NASA could use a smaller rocket, the Saturn V, instead of the giant Nova. Apollo's two spacecraft would be stacked in the same Saturn V, separate, and dock twice during the mission.

Grumman, previously a developer of innovative naval aircraft, would now be the prime contractor to build the Lunar Module, or LM. As the first human spacecraft ever built to fly only in a complete vacuum, it would need no wings or control surfaces, but it would need to navigate successfully to the lunar surface, keep its crew of two alive on the Moon, and blast back into lunar orbit, where one member of the crew would wait, tending to the command ship that would bring all three of them home. The craft would look more like a bug than a jet, with thin walls and a computer that previously took up the floor of a building. Like the Apollo command ship, which would comprise a crew compartment and an engine stage, the LM, too, would have two stages: a descent stage with an engine, fuel tanks, and landing legs, and a smaller ascent stage, housing the crew compartment and a smaller engine. Anything not required for the astronauts' safe return would be jettisoned the moment it was no longer needed: by shedding excess mass, Apollo would become lighter and lighter and easier to fly, until only the conical Command Module returned to Earth, with three men aboard, a bag of moon rocks, and several dozen rolls of photographic film (Brooks et al.).

An audacious plan, to be sure. But one, in 1961, that still left a thousand unanswered questions for a nation that had only ever sent one person into space for fifteen minutes. Could humans even survive the trip? Nobody knew for sure, and the only scientists who had studied the problem had their doubts. The space between the Earth and the Moon would offer an absence of air so complete that the body's moisture would evaporate straight through the skin. With temperatures shifting between two hundred degrees Fahrenheit in the Sun and two hundred below in shadows, exposed bodies would burn and freeze. Even if a craft could be fashioned to provide breathable air, moisture, and warmth, the weightlessness produced by the spacecraft's movements would disorient the body until it lost the ability to function.

So insurmountable were the problems of survival in space that at least a few scientists imagined that humans would never be able to travel to the Moon until humans somehow figured out how to re-engineer themselves to survive the lack of ambient pressure and oxygen, and huge temperature extremes (F. Freeman). To physiologists who pleaded for caution, NASA offered instead a confidence, bordering on faith, that the problems of survival in space would not require a new phase of human evolution. Humans in spacecraft and pressure suits would survive in space because they would have to if Americans were going to get to the Moon and back.

What would these suits look like? Simply putting humans in a rubber balloon offered no solution: rubber bladders expanded in space, and bending elbow and knee joints reduced the volume of rubber suits, increasing the pressure within them and making movement tiring. Engineers offered two solutions. The first, essentially a portable spaceship, would be a rigid container, either comprised of jointed segments like the body of an insect or a single large cylindrical container with flexible limb extensions, in which the astronaut would enjoy breathable air and some of the comforts of home, even on the Moon. Republic's voluminous Moon suit proposal featured a spacious compartment in which an astronaut could scratch his nose, enjoy a sandwich, or handle a camera; in the Moon's low gravity, the suit's weight would barely be an inconvenience. Though offering superior comfort, these suits did have one flaw: they were too big to fit inside the LM, let alone squeeze through its hatch.

As an alternative, and after nearly a decade of research and development, NASA seized upon a flexible suit that used a nylon restraint layer and specially designed joints and cable systems to maintain its internal volume even as the wearer moved and bent. The technology for such a suit, and the people who fabricated it, came from unlikely places, including the women's undergarment industry, which was already revolutionizing the management of weight and pressure using new synthetic materials. Built by the industrial parent of girdle-maker Playtex, the project Apollo spacesuit would set the standard for design, and modern variants remain in use today (Mallan).

Like many of the technologies used to land on the Moon, the spacesuit was completed not long before Apollo 11's flight and was modified throughout the Apollo lunar program to make it more effective. Protecting astronauts from the physical dangers of space ultimately proved, if not quite as easy as NASA had hoped, then at least easier than its physiologists had feared. What such a voyage might do to the human mind—the sudden "withdrawal of smoking, cokes, coffee, and snacks," the sudden "separation" of "familiar" social "supports"—was also unknown (Sells and Berry 168, 170, 172, 174–75). Psychologists feared the worst of the men who might fly into space, in part because they feared that anyone who wanted to fly into space was not stable to begin with. Popular culture throughout NASA's first decade raised fears of space madness: a mixture of fear, alienation, and religious mania that would await any person unlucky enough to live long enough in space to realize exactly where he was and what he had done. In TV shows from the period, like the *Twilight Zone*, and popular movies like *Marooned*, astronauts confront the prospect of their own death with a mixture of stoicism and complete mental breakdown, a fate that psychologists feared awaited space crews bound for mind-bending visits to the Moon (Wolfe; Sturges; Matheson; Heyes). In reality, however, heightened emotions in space capsules were exceedingly rare. In scouring the nation's test pilot schools for experienced aviators, NASA instead found men whose technical skills were matched by their self-control, like the fictional Major Nelson from *I Dream of Jeannie*, a TV series for which NASA vetted scripts to ensure technical accuracy (Cox 58–61).

Three such men, test pilots who had volunteered to become astronauts and been selected by NASA after a rigorous set of evaluations—one in 1962 and two in 1963—would become the crew of Apollo 11. Like the other pilots who joined them and many more against whom they competed, they were, by all appearances, intelligent, educated, and stable: confident in themselves and the space program (Korchin and Ruff 204–7; Santy). They were also all white men, due less to overt discrimination than to structural barriers that prevented women and minorities from acquiring the essential skills required for selection, including credentials from military test pilot schools (Weitekamp; Kevles). Unlike America's first astronauts of 1959, the Apollo 11 astronauts had joined the space program from promising careers in aviation knowing exactly what risks lay ahead of them, and with the knowledge that if successful, they would go to the Moon. Who would get to fly and when was determined years before flight. Why those men were chosen and not others was one of the closely guarded secrets of the space program and known only to two senior astronauts who went to their deaths never telling anyone why.

Who were the crew members of Apollo 11? Neither daredevils nor fools, they had turned the management of risk into a profession by carefully calibrating their emotions, but they were still human—so human, that their achievement appears all the more miraculous. Neil Armstrong was a former naval aviator whose career as a civilian test pilot for NASA had hit a rough patch after his daughter's death from leukemia (Hansen). Air Force pilot Michael Collins had joined NASA in 1963 after an unsuccessful attempt the year before and without any encouragement to do so, haunted by a lingering fear of his own inadequacy (Collins). The Air Force's Edwin "Buzz" Aldrin had a doctorate from MIT, a family history of depression and suicide, and, eventually, a growing problem with fidelity and alcohol (Aldrin and Warga). Like many of their peers, they were human beings who gave the better parts of themselves to their difficult, time-consuming, and dangerous work, often leaving troubled marriages in their wake. They lived in Houston, Texas, and commuted to sites around the country in jets they flew themselves, often with tragic results. And before they could go to the Moon, they would need to learn how to do everything such a mission would require.

For all three men, the opportunity came through NASA's Project Gemini: an enlargement of NASA's original Mercury spacecraft, offering room for a second crewmember and more robust navigation and control systems (Hacker and Grimwood). Over a period of twenty months from 1965 through 1966, ten crews developed the techniques they would need to get to the Moon, from navigating in space to rendezvousing with other spacecraft and leaving their vehicles to walk in space. Each offered obstacles to overcome. To fly in space required tossing out much of what these pilots understood about flight. The laws of orbital mechanics produced counterintuitive responses to the astronauts' efforts to control their movements: touching an object made it fly away; firing a thruster forward made the spacecraft climb and slow down. To perform the delicate calculations required to dock a craft in space, Gemini vehicles used the first onboard computers on spacecraft; only Buzz Aldrin had the mental acuity to fly without one.

Upon leaving the spacecraft, the astronauts' problems intensified. Venturing into the weightless vacuum of Earth's orbit, astronauts could do little more than stiffen their bodies and float: grabbing objects and moving from place to place created unintended movements that were difficult to control. As heat built up in spacesuits, astronauts became dehydrated, and their pulses soared. Walking in space and on the Moon would require not only new suit technology (especially liquid cooling) but new ways of controlling one's body. One answer to these problems, advocated by Aldrin, was the use of swimming pools to mimic the floating sensations of space.

It was in this period that American space capabilities began to outstrip Soviet ones. By the end of Project Gemini, the essential skills required of astronauts for their trip to the Moon had been mastered, despite numerous close calls and near misses: rocket engines that would not start, valves that would not seal, and hatches that would not shut. At various points, problems were so severe that they endangered the lives of the crew members, as on Gemini 8 when Neil Armstrong and David Scott's spacecraft began to tumble in orbit and the two men nearly blacked out (Hacker and Grimwood). Before the Moon Race had ended, space work had killed eight astronauts: four in flying accidents, three in fires, and one in a car crash. What we know of Soviet personnel losses, in space and in training and launch estimates, were ten times greater. Working for NASA was about as dangerous as flying in combat, but astronauts, if anything, craved trouble: resolving in-flight mishaps, Apollo 10 and 17 Astronaut "Gene" Cernan suggested, would prove their worth ("I almost dared her to quit on me," Cernan later recounted of his spacecraft). Such heroism might also, perhaps, absolve them of their guilt—while astronauts earned praise and fame, Cernan noted, his flight school classmates were flying the skies over Vietnam, taking similar losses in relative anonymity (Sington). Apollo 8 Astronauts Frank Borman and "Bill" Anders shared similar thoughts (Borman et al.).

None of Gemini's close calls, though, would match the challenge of landing on the Moon, a place so seemingly inhospitable that scientists had little idea if a craft could even successfully land on its surface. Was the Moon covered with boulders? Craters? A fine sand into which a lander would sink? Telescopes could not provide the answers, and science fiction provided a host of threatening environments more likely to chew up a lunar lander than provide sanctuary. Obtaining close imagery of the Moon was a priority for NASA's scientific community, as was landing an unpiloted probe on its surface to verify that humans could follow.

Throughout the mid-1960s, both the United States and the Soviet Union undertook efforts to acquire this essential information, vying as well to score major "firsts" in the annals of space exploration: photographing the Moon, crashing into it, and safely landing unpiloted probes on its surface. In almost all of these contests, the Soviet Union achieved earlier victories, with more capable American craft following shortly behind. This pattern injected dread into the minds of American space planners: while NASA, in the words of one astronaut, preferred to be right rather than first, politicians, journalists, and much of the public saw space exploration as a race that the US could not lose. Although NASA's capabilities in human spaceflight were robust and growing, the US could not be sure of winning the

Moon Race until it landed on the Moon and found no Russian cosmonauts already there.

With flights of its Ranger, Surveyor, and Lunar Orbiter craft (a repurposed American spy satellite), NASA verified that parts of the Moon would be flat enough and solid enough to land, and began to mark potential landing sites offering both good odds of a safe landing and areas of geological interest. For scientists, the Moon was a time capsule of Earth's formation: there, astronauts would find the same primordial rocks from which Earth had formed, unweathered by four-and-a-half billion years of water, air, and more recently, life. Most sites lay near the Moon's equator, offering easier navigation from Earth. Landing on the Moon's uneven surface, though, would still be difficult, likely testing the skills of even the best pilots.

In the LM, a craft so small it had neither seats nor bunks, standing astronauts would find control sticks, instrumentation, and tiny windows. During most of the LM's descent, the astronauts would not see the landing site; once they pitched over, the Moon's horizon would come into view, and they would need to manage a precious reserve of fuel to land, knowing that their propellant tanks gave them only one chance at a descent. This was, simply, work too complex for humans to master. In simulations, no astronaut could land the LM entirely manually; Apollo 11's computers would handle most of the heavy calculation and systems control, with astronauts monitoring, adjusting, and taking over limited manual control in the event of trouble.

The men and women who built, programmed, and troubleshot these computers, like the backup crews who made and revised checklists, were the largely unheralded extra crew members aboard Apollo 11 (Mindell); though Neil, Mike, and Buzz carried the ultimate burden and responsibility for the mission's success—Mike in Command Module *Columbia* and Neil and Buzz in Lunar Module *Eagle*—their voyage was a piece of a larger whole. Apollo 10, launched months before Apollo 11, provided NASA with a dress rehearsal of the landing. In Command Module *Charlie Brown* and Lunar Module *Snoopy*, astronauts orbited the Moon and practiced the descent, aborting before landing to verify that Apollo 11's crew would be able to return safely. A favorite of the astronauts and the public, Snoopy and fellow *Peanuts* characters by cartoonist Charles Schulz became official mascots of Project Apollo, utilized both to increase popular enthusiasm for spaceflight and to create visual reminders of the need to safeguard the lives of the astronauts through rigorous attention to detail, specifically with regard to crew safety.

While the LM could land virtually automatically, it had no ability to judge the flatness of terrain or detect obstacles like boulders. In order to ensure success, the LM still needed human pilots who might need to assume more control of the craft as it approached the surface. Practicing such flights proved so difficult that NASA tried to qualify its astronauts using a variety of exotic vertical take-off aircraft, including helicopters and a device made by Bell Aerosystems nicknamed the "flying bedstead," which used a jet engine to simulate the rocket engine of the LM. These training methods were controversial even then, with crashes frequent and ejection from the craft necessary to prevent the loss of the pilot.

Ultimately, all of the computer systems and all of the astronauts' training would be necessary to land on the Moon. Though relatively smooth in comparison to other flights, Apollo 11 experienced difficulties throughout the mission. A problem during the *Eagle*'s undocking from *Columbia* blew the lander so far off course that when *Eagle* approached the lunar surface, the crew had no idea where they were. On the descent, electronic resonances in the LM's radar circuits generated a torrent of false data that taxed the computer almost to its limits, causing the software to shut down nonessential systems and spit out alarms faster than ground controllers could decode them. Settling down in the last few hundred feet, Neil and Buzz, with their computer working to its limits and an unfamiliar field of craters and boulders lying ahead of them, arrested the programmed landing sequence and skidded across the lunar surface in search of flatter ground, settling on a quiet patch with seconds of fuel remaining (Brooks et al. 343–44; Mindell; Chaikin).

Landing just after 4 pm eastern time on July 20, 1969, the crew of *Eagle* could scarcely contain their enthusiasm. Though planning to sleep upon their arrival and step out of *Eagle* in time for morning television coverage, Armstrong and Aldrin accelerated their timetable with an 11 pm performance that captivated a global audience of about 530 million people. Stepping onto the lunar surface, they found fine lunar soil and a horizon free of haze and fog, making the judging of distance almost impossible. The Moon was lifeless and gray but possessed of a stark beauty that the astronauts struggled to capture in photographs. Too busy to gaze, they set to work, setting up experiments and collecting samples for their return to Earth (see fig 1.2) (Brooks et al. 345; Chaikin).

Figure 1.2. In this 1969 photograph by Apollo 11 Commander Neil Armstrong, Lunar Module Pilot Edwin "Buzz" Aldrin stands on the Moon's surface near the Solar-Wind Composition Experiment (NASA photo).

The planting of the American flag by the Apollo 11 crew during their two-hour moonwalk was somewhat less memorable than we might imagine. The astronauts had a busy schedule and talked little during the event. The lunar ground was too hard to hammer the flagpole to the proper depth, and the flag probably fell down when the astronauts blasted off the surface (Jones). When in space and on the Moon, many astronauts felt, rather than an all-consuming patriotism, a belief that for their voyages to have meaning, they would have to represent the achievements of humanity in addition to those of one country (Sington). They thought of their families and their personal struggles ("Charlie" Duke left a photograph on the lunar surface of himself with his wife and two sons) (Orwig), while some stewed over physical discomforts (John Young flatulated constantly from the oxygen dissolved in his water supply) (Smith 218). Looking back at the Earth while farting uncontrollably, most saw not competing nation-states, but a rare burst of color against an inky blackness that made political differences seem small in comparison (White; Borenstein). Plus, the Apollo astronauts had a larger worry on their minds.

Though itself a challenge, landing on the Moon was the easy part. Lifting off and docking with *Columbia*—literally hitting a bullet with another bullet—was the real difficulty Apollo 11 astronauts would encounter on the mission. Collins, circling overhead in *Columbia*, scribbled as much on the back page of one of his flight manuals (Collins). Fairly certain that he would survive no matter what, he secretly feared that Armstrong and Aldrin were on a one-way trip and would never leave the Moon alive (Sington). Indeed, launching rockets from Earth was still in its infancy, and in 1969, launching one from the surface of the Moon had never been attempted, let alone a rocket carrying people.

No film of Apollo 11's ascent exists, but later efforts to film this event using a remote television camera finally succeeded on Apollo 17. Photographs of *Eagle*'s ascent stage from *Columbia* show the fragility of this unusual flying machine. The skin of the craft, equivalent to three layers of aluminum foil, crumpled under the stress of launch from the lunar surface (see fig 1.3). With little margin for error, *Eagle* rose to meet *Columbia*, the two craft matching speed and altitude and docking successfully after a final moment of navigation trouble. Its mission over, *Eagle* was abandoned. Three days later, *Columbia* plunged through Earth's atmosphere, protected by a heat shield that charred and vaporized as the plasma consumed it. The cylindrical Service Module, lacking such protection, burned up and fell into the ocean, the last piece of a 360-foot-tall rocket consumed almost completely during Apollo 11's mission to the Moon.

Dangling from three large parachutes, *Columbia* settled into the waters of the Atlantic Ocean as a naval recovery fleet waited nearby. Ironically, Apollo's Command Module was far better at holding air in than keeping water out, and while it floated when intact, water often splashed through various ports, and drowning in the craft was a constant fear of the crew's. Once located by frogmen, the astronauts received biological isolation suits, intended not to protect them from Earth, but to contain whatever unknown microorganisms they might have unwittingly brought back from the Moon. The fear, stoked by Michael Crichton's *The Andromeda Strain*, later proved

Figure 1.3. The ascent stage of the Apollo 17 Lunar Module *Challenger* coasts in lunar orbit, photographed from Command Module *America* in 1972 (NASA photo).

unfounded: after a few weeks in quarantine, NASA determined that the Moon was as sterile and lifeless as it appeared (Crichton). The only life to be found upon it would have to come from Earth.

The urgency with which humans voyaged to the Moon in 1969 was never about what they would find there: "magnificent desolation," according to Buzz Aldrin (Jones). Rather, it was the conviction that the journey itself would change humanity for the better. Austerity, discomfort, and disappointment were all to be expected, but in the final moments of distance and alienation, something close to the true nature of humanity would be revealed . . . perhaps a new appreciation for planet Earth, or, in the words of the US Army missile program's General John Medaris, "a new understanding of man's relationship with the infinity of Divine Creation" (De Groot 93). Like the children heading out into the night in Robert Frost's

"Going for Water," the Apollo 11 astronauts did not seek sustenance from the Moon's distant and imposing surface, but they knew that the rewards bestowed upon humanity for completing the journey would quench their thirst (Frost 38):

THE WELL was dry beside the door,
And so we went with pail and can
Across the fields behind the house
To seek the brook if still it ran;

We ran as if to meet the moon
That slowly dawned behind the trees,
The barren boughs without the leaves,
Without the birds, without the breeze.

But once within the wood, we paused
Like gnomes that hid us from the moon,
Ready to run to hiding new
With laughter when she found us soon.

Works Cited

Aldrin, E. E., and W. Warga. *Return to Earth*. Random House, 1973.

Baxter, James Phinney. *Scientists against Time*. Little, Brown and Company, 1946.

Bilstein, Roger E. *Flight in America: From the Wrights to the Astronauts*, 3rd ed. Johns Hopkins UP, 2001.

———. *Stages to Saturn: A Technological History of the Apollo/Saturn Launch Vehicles*. Scientific and Technical Information Branch, National Aeronautics & Space Administration, 1980.

Borenstein, Seth. "Capturing the Big Picture on Earth Day." *Los Angeles Times*, 22 Apr. 2007, pp. A20–21.

Borman, F., et al. "John H. Glenn Lecture: An Evening with the Apollo 8 Astronauts." YouTube. Uploaded by *Smithsonian Institution, National Air and Space Administration*, 18 Dec. 2013. https://youtu.be/Q2h_FtLzrrU.

Brooks, Courtney G., et al. *Chariots for Apollo: A History of Manned Lunar Spacecraft*. National Aeronautics and Space Administration, 1979.

Chaikin, Andrew. *A Man on the Moon: The Voyages of the Apollo Astronauts*. Viking, 1994.

Collins, Michael. *Carrying the Fire: An Astronaut's Journeys*. Farrar, 1974.

———. "Ye Ole Lunar Scratch Pad." *Outside the Spacecraft*, https://airandspace.si.edu/exhibitions/outside-the-spacecraft/online/image-detail.cfm?id=9804. Accessed 19 Jan. 2022.

Cox, S. *Dreaming of Jeannie: TV's Prime Time in a Bottle*. St. Martin's Griffin, 2000.

Crichton, Michael. *The Andromeda Strain*. Random House, 1969.

De Groot, Gerard J. *Dark Side of the Moon: The Magnificent Madness of the American Lunar Quest*. New York UP, 2006.

Freeman, Fred. "Man Remade to Live in Space." *LIFE*, Jul. 1960, p. 77.

Freeman, Mae Blacker. *You Will Go to the Moon*. Beginner Books, 1959.

French, F., and C. Burgess. *In the Shadow of the Moon: A Challenging Journey to Tranquility, 1965–1969*. University of Nebraska Press, 2007.

Frost, Robert. *A Boy's Will*. Henry Holt and Company, 1915.

Hacker, Barton C., and James M. Grimwood. *On the Shoulders of Titans: A History of Project Gemini*. National Aeronautics and Space Administration, 1978.

Hansen, J. R. *First Man: The Life of Neil A. Armstrong*. Simon & Schuster, 2005.

Heyes, Douglas, scr. "And When the Sky Was Opened." *The Twilight Zone*. 11 Dec. 1959.

———. *In the Shadow of the Moon*. Directed by David Sington, Discovery Films, 2007.

Johnson, Stephen B. *The Secret of Apollo: Systems Management in American and European Space Programs*. Johns Hopkins UP, 2002.

Jones, Eric M. "One Small Step." *Apollo 11 Lunar Surface Journal*, 1995, https://www.hq.nasa.gov/alsj/a11/a11.step.html.

Kennedy, John F. "Excerpts from 'Urgent National Needs,' Speech to a Joint Session of Congress, May 25, 1961." *Exploring the Unknown: Selected Documents in the History of the U.S. Civil Space Program*, edited by John M. Logsdon, National Aeronautics and Space Administration, 1995, pp. 453–54.

Kepler, Johannes, and Edward Rosen. *Somnium: The Dream, or Posthumous Work on Lunar Astronomy*. University of Wisconsin Press, 1967.

Kevles, Bettyann H. *Almost Heaven: The Story of Women in Space*. Basic Books, 2003.

Korchin, Sheldon J., and George E. Ruff. "Personality Characteristics of the Mercury Astronauts." *The Threat of Impending Disaster, Contributions to the Psychology of Stress*, edited by George H. Grosser et al., MIT Press, 1964, pp. 197–207.

Lambright, W. H. *Powering Apollo: James E. Webb of NASA*. John Hopkins UP, 1995.

Launius, Roger D. *NASA: A History of the U.S. Civil Space Program*. Krieger Pub. Co., 1994.

Lehman, M. *This High Man: The Life of Robert H. Goddard*. Farrar, 1963.

Mallan, L. *Suiting up for Space: The Evolution of the Space Suit*. John Day Co., 1971.

———. *Marooned*. Directed by John Sturges, Columbia Pictures Corporation, 1969.

Matheson, Richard, scr. "Death Ship." *The Twilight Zone*. 1963.

McDougall, Walter A. . . . *the Heavens and the Earth: A Political History of the Space Age*. Johns Hopkins UP, 1997.

Mindell, David A. *Digital Apollo: Human and Machine in Spaceflight*. MIT Press, 2008.

Neufeld, Michael J. *Von Braun: Dreamer of Space, Engineer of War*. A.A. Knopf, 2007.

Newton, Isaac, et al. *Philosophia Naturalis Principia Mathematica*. Jussu Societatis Regiae ac typis Iosephi Streater, prostat apud plures bibliopolas, 1687. *HathiTrust*, https://catalog.hathitrust.org/Record/000383175.

Orwig, Jessica. "There's a Hidden Message on This Family Portrait That an Astronaut Left on the Moon." *The Independent* (UK), 2 Nov. 2015. https://www.independent.

co.uk/news/science/apollo-16-astronaut-explains-hidden-message-behind-the-family-portrait-he-left-on-the-moon-a6718111.html.

"Rocket to the Moon: Man May Travel to Earth's Satellite within 25 Years." *LIFE*, Jan. 1949, pp. 67–73.

Santy, Patricia A. *Choosing the Right Stuff: The Psychological Selection of Astronauts and Cosmonauts*. Praeger, 1994.

Sells, S. B., and C. A. Berry. "Human Requirements for Space Flight." *Man in Space: The United States Air Force Program for Developing the Spacecraft Crew*, edited by K.F. Gantz, Duell, 1959, pp. 161–77.

Serviss, Garrett P. *Edison's Conquest of Mars*. Carcosa House, 1947.

Siddiqi, Asif A. *Challenge to Apollo: The Soviet Union and the Space Race, 1945–1974*. National Aeronautics and Space Administration, NASA History Division, Office of Policy and Plans, 2000.

Smith, Andrew. *Moondust: In Search of the Men Who Fell to Earth*. Fourth Estate, 2005.

Swenson, Jr., Loyd S., et al. *This New Ocean: A History of Project Mercury*. Scientific and Technical Information Division, Office of Technology Utilization, National Aeronautics and Space Administration, 1966.

Webb, James E. *Space Age Management: The Large-Scale Approach*. McGraw-Hill, 1969.

Weitekamp, M. A. *Right Stuff, Wrong Sex: America's First Women in the Space Program*. Johns Hopkins UP, 2004.

Westfall, Richard S. *Never at Rest: A Biography of Isaac Newton*. Cambridge UP, 1983.

White, Frank. *The Overview Effect: Space Exploration and Human Evolution*. American Institute of Aeronautics and Astronautics, 1998.

Wolfe, P. *In the Zone: The Twilight World of Rod Serling*. Bowling Green State University Popular Press, 1997.

2

The Hidden Chemistry behind the Apollo 11 Mission

SABRINA G. SOBEL

> I think we're going to the Moon because it's in the nature of the human being to face challenges. It's by the nature of his deep inner soul. We're required to do these things just as salmon swim upstream.
>
> Neil Armstrong ("Quotes")

Everyone thinks that rocket science is all about physics, but few people realize just how much chemistry went into making Apollo 11 possible. This chapter focuses on the behind-the-scenes contributions that chemistry made to the success of the Apollo missions and how these innovations later made their way into our everyday lives. Chemistry impacted the choice of materials used for everything from rocket engines to spacesuits, playing an essential role in both obvious and more hidden areas. One obvious chemical contribution was the development of different rocket fuels, but the hidden chemistry was in the production, storage, and use of fuels such as liquid oxygen and hydrogen. Polymers such as nylon, Dacron, and Mylar were key to the design of a functional spacesuit, and carbon dioxide scrubbers were used for air handling in the closed system of a spacesuit. Hidden chemistry not only brought men to the Moon; today it quietly continues to influence our daily lives.

The Apollo program started as an ambitious if uncertain effort that eventually succeeded in sending humans to the Moon and returning them safely. Astronauts knew the dangers inherent in this endeavor and thought it would be a miracle if they were successful, but this did not stop them. As President John F. Kennedy implied, such striving is deeply embedded in human nature: "We choose to go to the Moon in this decade and do the other things, not because they are easy but because they are hard." To do the seemingly impossible, the Apollo program catalyzed the integration of diverse scientific and technological fields, all to achieve the same goals: transporting humans to the moon, facilitating their work on the Moon, and returning them safely to Earth. The men and women of the Apollo program were attempting the previously unimaginable: escaping Earth's gravity, maneuvering to another orbiting body (the Moon), carrying with them everything they needed to survive—food, air, fuel to return—maneuvering back to the Earth and successfully returning without burning up in the atmosphere on re-entry.

A Note on Sources

Because Apollo was a government program, there is a wealth of information publicly available so that anyone can do a deep dive into whatever particular aspect of the program is of interest. The areas that caught my curiosity were rocket fuel and fuel handling, the design and performance of spacesuits, and enabling the respiration of humans in space. These are all technical issues that required knowledge of chemistry applied in novel ways. That is why I refer to these technological feats as the "hidden chemistry" of the Apollo program. My greatest resource was the National Aeronautics and Space Administration (NASA) history website, https://history.nasa.gov/, in which I found a thorough report: SP-368 Biomedical Results of Apollo, Section VI, that covers space meals, potable water, waste management, bioinstrumentation, environmental control systems, and the spacesuit (Extravehicular Mobility Unit) (Parker Jr.). To get a sense of the rocket fuel challenges and implementation, I turned to the aeronautics section on the same site as well as two infographics created by Karl Tate. A journey through this or any portion of the history of the NASA Apollo program will provide the reader with an enhanced appreciation of the amazing nature of the Apollo missions and a renewed appreciation of how society has benefited from the chemistry and other technologies that were refined in service of the effort to put men on the Moon.

The Hidden Chemistry of Rocket Fuel

> As an astronaut, especially during launch, half of the risk of a six-month flight is in the first nine minutes.
>
> Chris Hadfield (Foster)

One of the central issues that NASA engineers faced was propelling a rocket out of Earth's gravity. It is the most dangerous part of a space mission, as pointed out by Astronaut Dr. Chris Hadfield (Foster). The search for efficient, yet powerful rocket fuels started long before the Apollo mission and involved some daring and unusual characters, such as John Parsons, a founding member of the Jet Propulsion Laboratory who later died in a mysterious explosion (Pendle). The sheer scale of this endeavor really tested the NASA engineering team's ability to handle enormous quantities of highly unstable and flammable chemicals. For the Saturn V rocket, engineers settled on a kerosene-liquid oxygen system to lift the rocket into space (fig. 2.1). They had worked with a kerosene-oxygen rocket design previously, so this was "just" a matter of scaling up known technology.

To escape Earth's gravity, the Saturn V rocket needed to achieve a velocity of 20,000 mph (Kurtus). Stage S-1C (fig. 2.1), the one that burned 203,400 gallons of kerosene with 318,000 gallons of liquid oxygen, generated 7.5 million pounds of thrust in 2.5 minutes, and accelerated the rocket to about 6,160 mph, only 31 percent of the necessary escape velocity. This astronomical amount of fuel got the rocket only one-third of the way to leaving Earth. In addition, having flammable fuel and so much oxygen stored in such proximity is an enormous hazard. Astronauts dying in an explosion at liftoff was a very present danger. Design of the engine cones to control thrust created by the combustion products, carbon dioxide and water vapor, was another technical challenge, well-described in the *NOVA* episode "Apollo's Daring Mission" (Wolfinger).

After Stage SC-1 was spent, liquid hydrogen (LH_2) and liquid oxygen (LOx) were combusted in two more rocket stages to create thrust from water vapor and thus achieve escape velocity and propel the astronauts to the Moon. Small attitudinal thrusters burning more LH_2 and LOx were used to guide the rocket in space. This was practical since both the LH_2 and LOx were stable in the cold, ambient temperatures of space. As cryogenic technology advanced, NASA moved to using just LH_2 and LOx in the Space Shuttle era (fig. 2.2), but not without its own tragedy—the Challenger

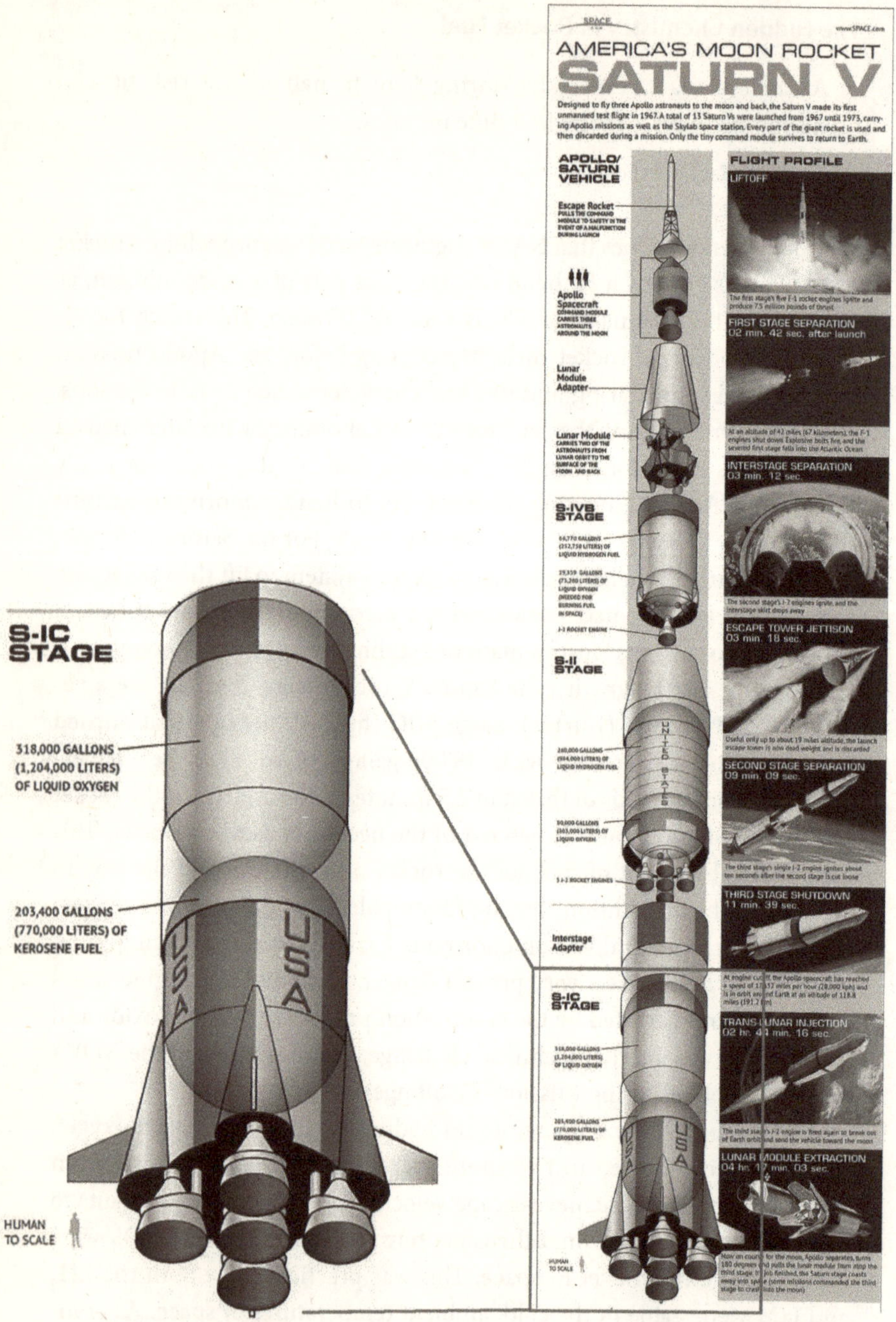

Figure 2.1. Diagram showing components of the Saturn V rocket and highlighting the first stage. Graphic by Karl Tate for Space.com.

Table 2.1. Technical specifications compared between the Saturn V rocket and the Space Shuttle (data from Tate). LOx = liquid oxygen = $O_{2(l)}$

	Saturn V	**Space Shuttle**
Payload	Apollo/Skylab	Shuttle Orbiter
Operation Years	1967–1973	1981–2011
Destination	Earth orbit, Moon	Earth orbit
Fuels	Kerosene (770,000 L), LH_2, LOx	Solid fuel, LH_2, LOx
Lift Capability	130 metric tons	24 metric tons
Thrust	7.5 million lbs (3.4 million Kg)	7.8 million lbs (3.54 million Kg)

disaster—caused by the failure of an O-ring on one of the solid rocket boosters that became brittle below freezing temperatures. Hot gasses from the solid rocket booster ruptured the liquid hydrogen fuel tank, leading to a complete structural failure of the liquid hydrogen and oxygen tanks and subsequent explosion of the fuel and destruction of the vehicle (Garber).

Looking at Table 2.1, one can see that an almost absurdly large amount of kerosene was needed to fuel the Saturn V as compared to our everyday use of kerosene in camp lanterns or home heating oil, which is a closely related hydrocarbon mix. This observation is connected with the question of why liquid oxygen was necessary for the combustion of kerosene in the rocket.

Earth's atmosphere is approximately 20 percent O_2—more than sufficient for the combustion of kerosene under ordinary circumstances. But how much oxygen is needed to combust over 200,000 gallons (770,000 L) of kerosene? Simple chemistry calculations can elucidate the issue. Kerosene consists of about 70 percent branched, straight chain and cycloalkanes, about 25 percent alkylbenzenes, and about 5 percent olefins, $C_{10}{-}_{16}H_{22}{-}_{36}$ which can be approximated as $C_{13}H_{28}$. The flash point of kerosene vapor is 37–65°C (100–150°F), ignition temperature is 220°C (428°F), and freezing point is −40°C (−40°F, 233 K) (Shepherd). Thus, kerosene is liable to blow up before ignition if conditions are not carefully controlled by minimizing the concentration of kerosene vapor before launch.

Kerosene is not practical for generating thrust in space since it will freeze. It is useful only for propelling the rocket into outer space but needs to be used in tightly controlled conditions. To give a sense of the scale of energy released, which is turned mostly into thrust (work), the burning of

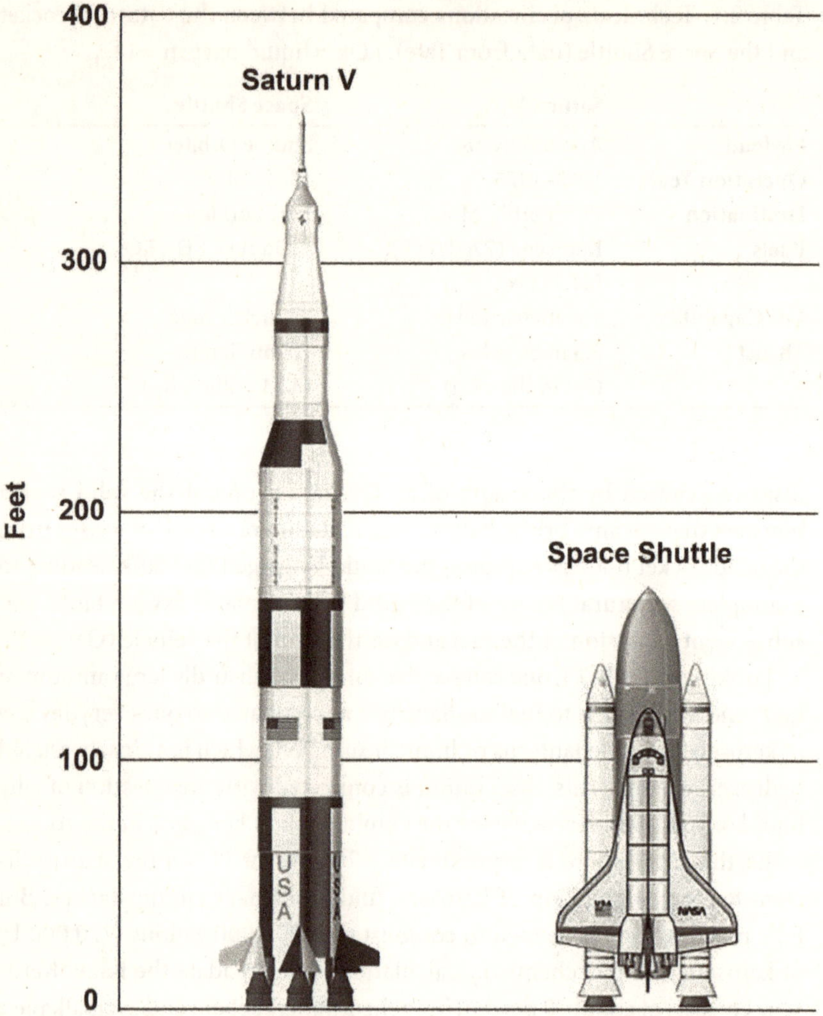

Figure 2.2. Graphic showing the Saturn V rocket compared to the Space Shuttle.

770,000 L of kerosene releases about 27,500,000 MJ, or enough energy to boil 32 swimming pools of water.

$$C_{13}H_{28} + 21\,O_2 \rightarrow 13\,CO_2 + 14\,H_2O \quad (1)$$

The sheer amount of kerosene to be burned (770,000 L or 203,500 gal) in a very short amount of time (2.5 minutes) requires a massive amount of oxygen (eqn. 1), approximately 3,300 kmoles oxygen or 8 million cubic meters of air, an impossible amount to expect to mix with the kerosene as it burns in the engines. Thus, the need for oxygen in a concentrated liquid form (LOx, $O_{2(l)}$) is made clear.

The 1.2 million liters of LOx in the S-1C stage provide 76,000 kmoles oxygen, which is more than enough oxygen available for mixing with the kerosene within the engines. However, handling liquid oxygen requires cryogenic technology since it only is liquid between 54K and 90K (-219°C to -183°C, -362°F to -297°F). That was yet another technical challenge for the Saturn V rocket engineers. They needed to construct a massive thermos to contain the LOx and insulated tubing to deliver it to the thrusters without having the LOx vaporize or freeze the nearby kerosene. In addition, having flammable fuel and so much oxygen stored in such proximity is an enormous fire and explosion hazard, which NASA learned during the development of the Saturn V rocket.

The Hidden Chemistry of Spacesuits

> Mr. McGuire: I want to say one word to you. Just one word.
> Benjamin: Yes, sir.
> Mr. McGuire: Are you listening?
> Benjamin: Yes, I am.
> Mr. McGuire: Plastics.
> Benjamin: Exactly how do you mean?
> Mr. McGuire: There's a great future in plastics. Think about it. Will you think about it?
>
> *The Graduate* (1967)

Imagine having to take everything you need to breathe with you just to walk out the door. Nowadays, with all of the news and movies portraying space travel, we don't have a hard time at all imagining this exercise. Back in the 1960s, the challenges of stepping out into space safely were not so familiar to people. Since the Apollo 11 mission included a Moon walk, it was imperative to develop a spacesuit aka Extravehicular Mobility Unit (fig. 2.3). Consider the technical challenges of designing such a suit. In order to keep the astronaut alive while allowing him to function, it needed to

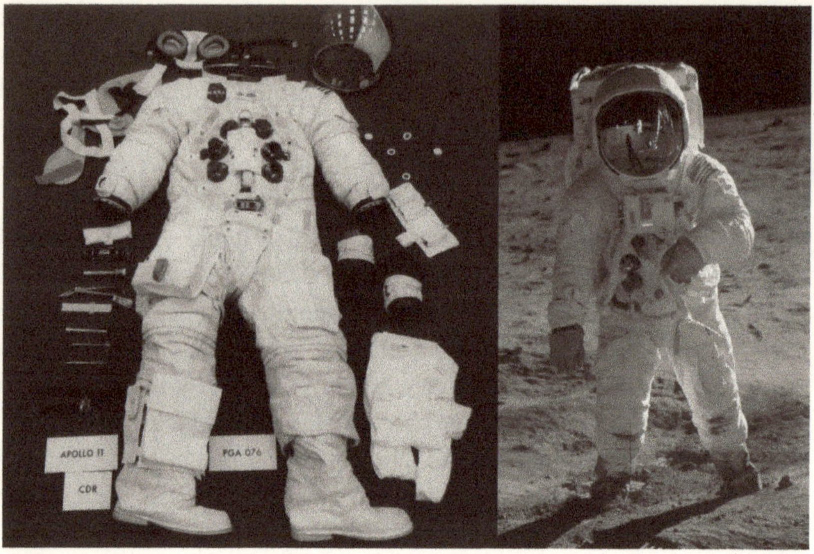

Figure 2.3. Pictures of the Apollo 11 spacesuit in parts and in action. Left image from "Neil Armstrong's Space Suit (EMU)," right image from "Buzz Aldrin on the Moon in 1969."

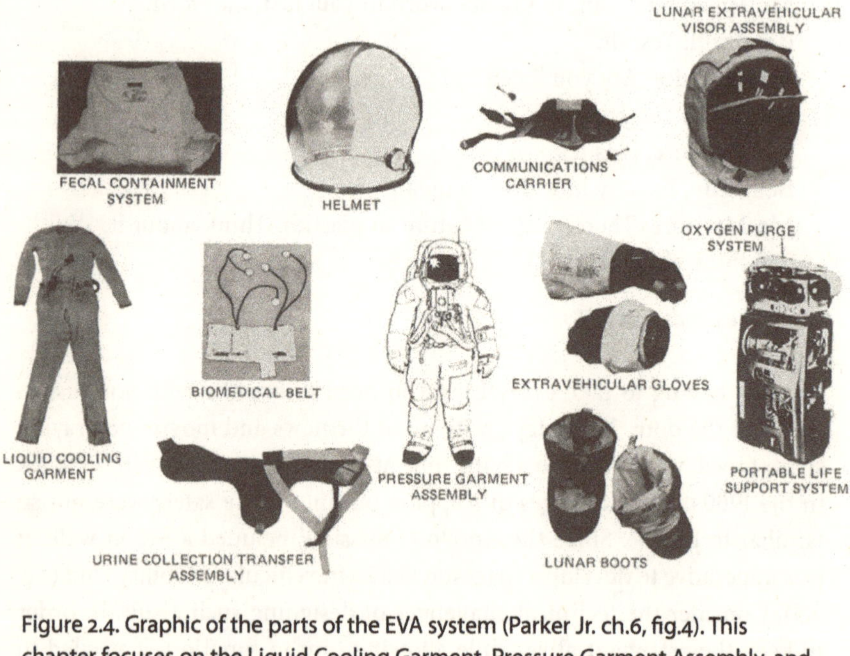

Figure 2.4. Graphic of the parts of the EVA system (Parker Jr. ch.6, fig.4). This chapter focuses on the Liquid Cooling Garment, Pressure Garment Assembly, and Portable Life Support System.

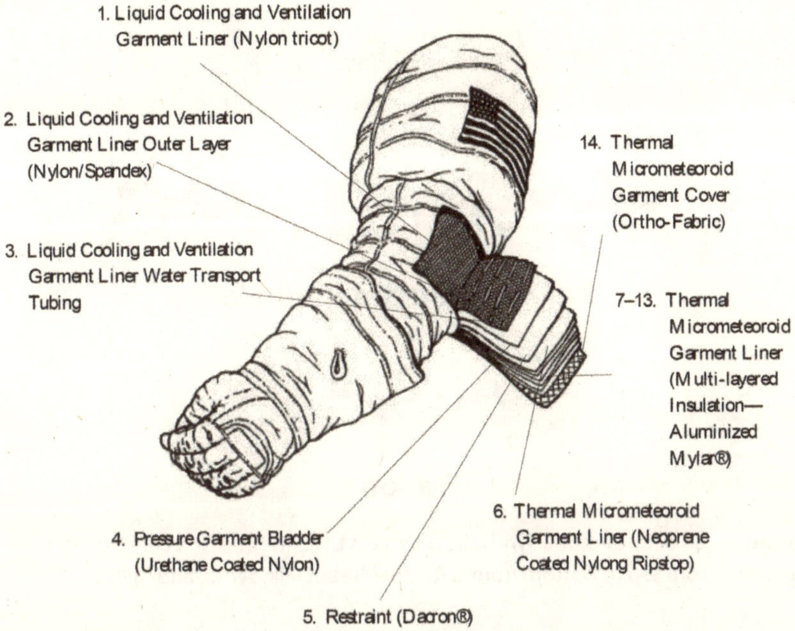

Figure 2.5. Details of the pressure garment assembly showing the different layers of material (Vogt 18).

be impermeable and pressurized, yet flexible and mobile. Equipment built into the suit had to enable gas handling and exchange for breathing, carry oxygen sufficient for a jaunt on the Moon, have thermal control so that the astronaut didn't freeze or bake, humidity control, and personal waste management.

Plastics, i.e., Polymers

As alluded to in the movie *The Graduate,* which came out two years before Apollo 11, the solution to fabricating a spacesuit that could accomplish all of these tasks was to use plastics (synthetic polymers composed of linked strands of carbon atoms) and elastomers (polymers with elastic properties). The pressure garment assembly (outer layer, figs. 2.4 and 2.5) was constructed of multiple layers of Dacron, nylon, aluminized Mylar, neoprene, and urethane. The Dacron and nylon acted as a flexible scaffold for the urethane and neoprene, rubbery polymers that make the woven Dacron and nylon impermeable without losing flexibility (Jones). Nylon was invented

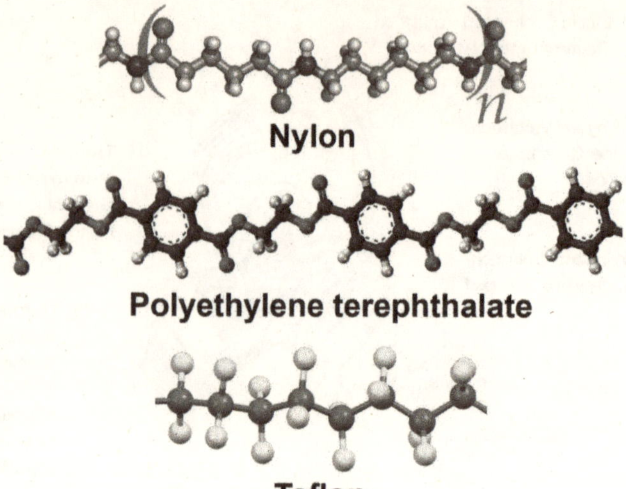

Figure 2.6. Structures of the synthetic polymers used in spacesuit construction (graphics from top to bottom from GYassineMrabetTalk, Jynto, and Mills).

in 1935, Dacron (spun/woven polyethylene terephthalate) in 1945, neoprene in 1930, and urethane in 1937—all products of the great twentieth-century flowering of synthetic organic chemistry (Knight). Nylon is a polyamide composed of a diamine and a dicarboxylate. The length of the linkers (connections between polymer strands) can be varied to change the properties of the resultant polyester. For example, nylon-6,6 is made from 1,6-hexane dicarboxylic acid polymerized with 1,6-hexanediamine, both of which have a 6-carbon backbone, thus "6,6" in the nylon name. Dacron is a commercial name for a woven version of polyethylene terephthalate, a polyester composed of ethylene glycol polymerized with 1,4-phthalic acid (terephthalic acid).

Teflon is a polymer of tetrafluoroethylene, C_2F_4 (fig. 2.6, bottom), that was discovered serendipitously when a gas cylinder full of C_2F_4 gas spontaneously polymerized. Because of all the fluorines, Teflon is very unreactive, which is why it is non-stick for foods.

Aluminized Mylar, the everyday Mylar that is familiar to us, was manufactured by DuPont starting in the 1950s and is composed of multiple thin layers of polyethylene terephthalate and aluminum foil adhered together (fig. 2.6). The Mylar sheets are created by extruding the polymer

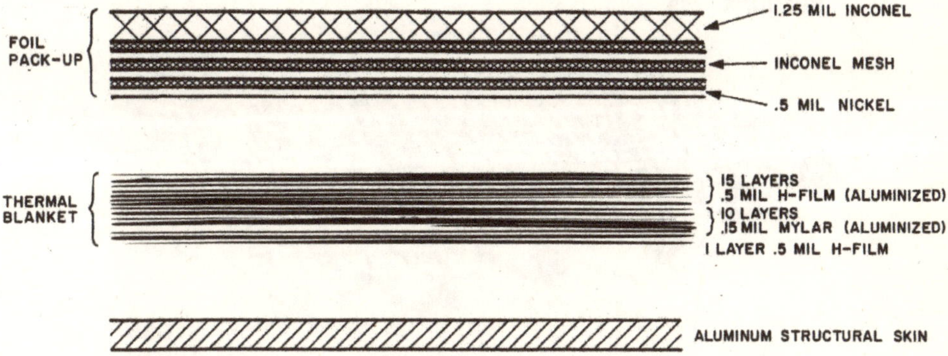

Figure 2.7. Detail of the Thermal Blanket layered construction (Cardin and Strasburger).

into thin sheets under constant tension, resulting in a strong, impermeable thin sheet that has a high tensile strength. The aluminum layers are vapor-deposited onto the thin Mylar sheets. As shown in fig. 2.7, each individual layer was 0.15–0.5 mm thick, and in the thermal blanket used in the Apollo 11 spacecraft, there were 25 aluminized Mylar film layers capped by a final plain Mylar layer. The aluminized Mylar was another impermeable but flexible layer used in the pressure garment assembly and thermal blankets. It also reflects infrared energy—heat waves—due to the layers of aluminum. Overall, Mylar provided superior thermal barrier protection along with fire and micrometeoroid protection for the duration of the EVAs on the lunar surface. The outer layer was "Beta cloth," woven fiberglass coated with Teflon, which also provided protection from micrometeoroid impacts that could punch microscopic holes in the spacesuit, while still being flexible, allowing movement. Since the spacesuit was impermeable and reflected thermal energy, Tygon tubing was woven in between nylon and spandex fabric (fig. 2.8) to create the Liquid Cooling Garment (fig. 2.4, middle left) to keep the astronaut from overheating.

The spacesuits for the astronauts had to be assembled by hand. For that, NASA turned to the International Latex Corporation in Delaware—a manufacturer of brassieres, thus the unforgettable article title: "Neil Armstrong's Spacesuit Was Made by a Bra Manufacturer" (Chaikin). Seamstresses at the International Latex Company had experience working with the stretchy fabrics and were capable of stitching the seams so evenly that

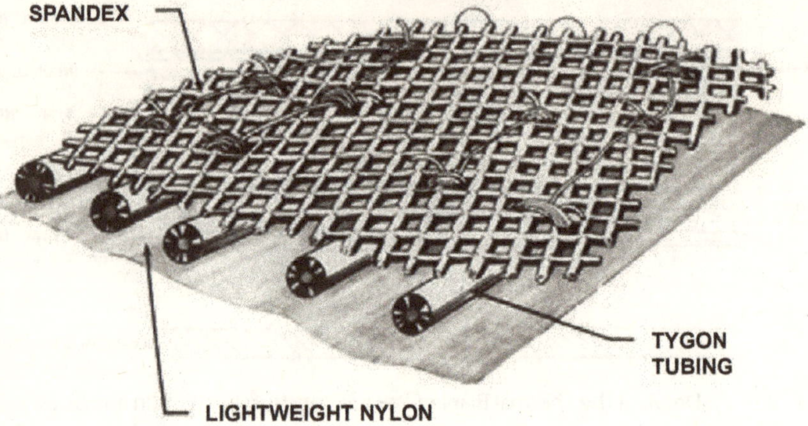

Figure 2.8. Detail of the construction of the Liquid Cooling Garment assembly (Parker Jr. ch.6, fig.15).

there were less than 1/32nd inch errors in the spacing of every stitch. Steel aircraft cables were sewn into the suit to provide structure to it so that it would remain moveable both inside and outside the Moon lander. This was important because as the astronaut stepped out into the near-vacuum on the Moon's surface, the internal pressure in the spacesuit would cause it to "blow up like a balloon" if not for a semi-rigid "skeleton" of aircraft cables (fig. 2.9).

The final consideration that goes into designing a spacesuit is that of "personal waste management" (i.e., urine and feces). The Fecal Containment System (fig. 2.4, top left) looks suspiciously like a diaper, so no surprise there. There was a separate Urine Collection Transfer Assembly (fig. 2.4, bottom left) that astronauts used as well. To minimize excess weight in the Apollo 11 rocket for the return trip, astronauts released urine out of Apollo 11 to outer space, where it immediately froze, leading to a great quote,

> Some astronauts describe the routine flushing of urine into space, where the freezing temperatures turn the droplets into a cloud of bright, drifting crystals, as being among the most amazing sights they saw on an entire voyage. (Cernan)

As illustrated by this observation from Astronaut Eugene Cernan, beauty can be found in the strangest circumstances.

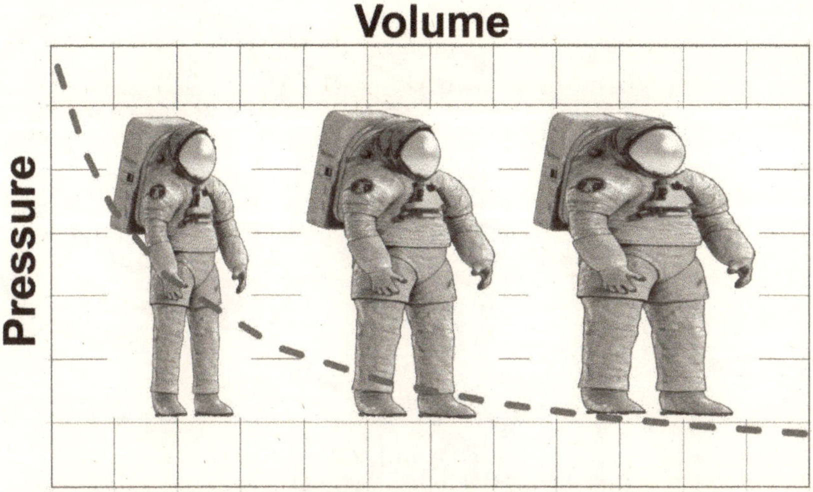

Figure 2.9. The "Astronaut's Dilemma"—as external pressure decreases, the constant internal pressure will cause a non-supported spacesuit to blow up and decrease free movement of the astronaut.

The Hidden Chemistry of Breathing

> There were many uncertainties about how well our Lunar module systems and our Pressure suit and backpack would match the engineering predictions in the hostile lunar environment. We were operating in a near-perfect vacuum with the temperature well above 200 degrees Fahrenheit with the local gravity only one-sixth that of Earth.
>
> Neil Armstrong ("Quotes")

Astronauts were able to carry their own air supply everywhere they went using the Portable Life Support System (fig. 2.10). When we breathe in, we absorb the oxygen from the air, and then when we breathe out, we expel carbon dioxide. The Portable Life Support System had tanks of breathable air (20 percent oxygen in nitrogen gas, like ordinary air), as well as an ingenious carbon dioxide absorbent system, to both avoid hypoxia (not enough oxygen) and minimize the volume of gases that needed to be carted around. Carbon dioxide gas needs to be removed from the air being breathed because otherwise carbon dioxide will displace oxygen and inhibit its absorption in the lungs, leading to suffocation. The carbon dioxide

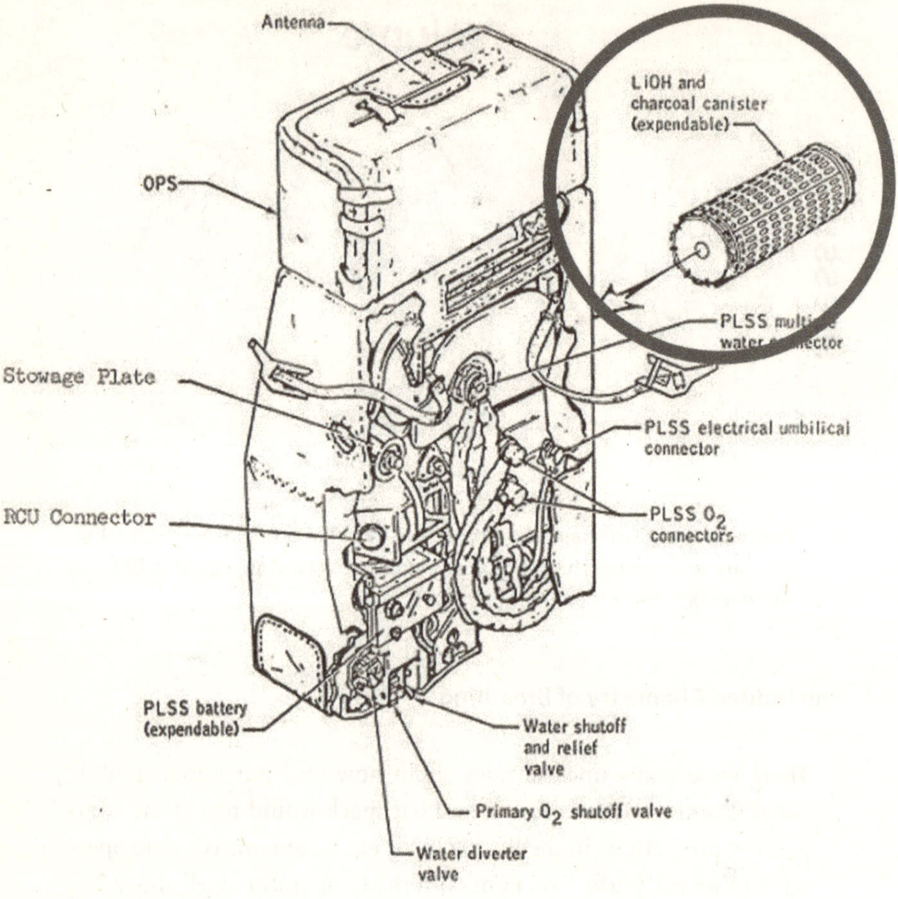

Figure 2.10. Close-up of the Portable Life Support System with the LiOH and charcoal canister circled (Vogt 27).

removal system consisted of a replaceable filter with a chemical carbon dioxide scrubber and charcoal to absorb odors. A replaceable canister meant that the Portable Life Support System could be used on more than one Moon walk as long as new oxygen tanks and a fresh carbon dioxide scrubber were installed.

Carbon Dioxide Scrubbers

Respiration, which supplies our cells with energy, involves the transformation of glucose and oxygen to carbon dioxide and water (eqn. 2). In a closed system, such as a spacesuit, carbon dioxide can build up and lead to hypoxia and ultimately suffocation.

$$C_6H_{12}0_6 + 6O_2 \rightarrow 6CO_2 + 6H_2O \quad (2)$$

The carbon dioxide removal system consisted of a replaceable filter with lithium hydroxide, to absorb the carbon dioxide, and charcoal, to absorb odors. Lithium hydroxide is a lightweight solid reactant that combines with carbon dioxide to form lithium carbonate, another solid (eqn. 3).

$$2LiOH_{(s)} + CO_{2(g)} \rightarrow Li_2CO_{3(s)} + H_2O_{(l)} \quad (3)$$

As with the lightweight lithium-ion batteries in our cell phones and laptop computers, NASA scientists turned to the periodic table of elements and chose the lightest of the alkali metals to use in the carbon dioxide absorbent to minimize weight. The alkali metals are in the first column of the periodic table of elements and contain familiar elements such as sodium (Na) and potassium (K), both of which are heavier than lithium (Li).

Practical Applications of Apollo's Hidden Chemistry

Tang—the orange powdered beverage with Vitamin C—was not the only innovation from the space program that made its way into our everyday lives. NASA scientists learned a lot about cryogenics—the science of handling large amounts of very cold liquids, such as liquid oxygen. Cryogenic technology is critical for Magnetic Resonance Imaging (MRI) machines at medical offices. The very strong magnets in an MRI machine work at liquid helium temperatures, which requires a two-level cryogenic thermos with liquid nitrogen in the outer jacket (77K, -320°F), and liquid helium in the inner jacket (4K, -450°F). Creating a commercial instrument with advanced cryogenics was not possible before extensive research into how to handle and store large volumes of very cold liquids, part of which was developed in the NASA Apollo program.

The layered multi-polymer construction of the spacesuit led to the development of fire-resistant protective clothing as well as thermal blankets and jackets. Aluminized Mylar layered with flexible outer polymers creates a durable material that reflects heat. For the thermal blanket or jacket, body heat is being trapped and reflected back to the person wearing the garment thanks to the aluminized Mylar. The opposite is true for the fire protection clothing, in which the aluminized Mylar layer is on the outside to reflect away the fire's heat. Helium balloons are made of aluminized Mylar to slow the loss of helium since both Mylar and the thin layer of aluminum are impermeable to helium. Normally, helium, a very light monoatomic gas, is small enough to slip through nanopores in most common rubber balloons. In contrast, the way that polyethylene terephthalate is extruded into Mylar results in an impermeable plastic sheet. Helium is only lost through the seams of a Mylar balloon.

Scuba rebreathers were conceived as a result of NASA Apollo science research. The idea that carbon dioxide can be trapped in a filter has been incorporated into specialized scuba equipment used by the military, underwater photographers, and marine biologists that reduces bubbles and noise and extends the life of the diver's air supply. In scuba rebreathers, air is recirculated instead of being discharged into the water. As in a spacesuit, carbon dioxide needs to be removed from the air stream to avoid hypoxia and suffocation. The chemistry was further refined to actually have the carbon dioxide trapping agent release oxygen as a result of the chemistry in the trapping process, which further extends the air supply. Solid potassium superoxide is loaded into the canister; it reacts with exhaled carbon dioxide to form potassium carbonate and free oxygen gas:

$$4KO_{2(s)} + 2CO_{2(g)} \rightarrow 2K_2CO_{3(s)} + 3O_{2(g)} \quad (4)$$

This is an efficient reaction since it takes a solid (KO_2) and a gas (CO_2) and transforms it into another solid (K_2CO_3) and gas (O_2). The manufacture of potassium superoxide is established, and KO_2 is stable as long as it is not exposed to air or water. However, it is a one-way reaction, so the carbon dioxide scrubber in the scuba rebreather needs to be swapped out for a fresh one when the potassium superoxide is used up.

Conclusion

It is illuminating to delve into some applications of chemistry in the Apollo space program. Normally we focus on the technology without stepping back and asking how the technology was developed using fundamental scientific knowledge from chemistry, biology, and physics. In the case of rocket propellant, new technology was created as a result of the necessary scale of the process, dealing with the problem of handling and rapidly combusting thousands of gallons of kerosene and liquid oxygen. For the spacesuit, multiple separate technologies were blended to create a unique result, an impermeable, flexible suit that carries its own air supply, thermal regulation, and waste systems. A constant political "push me-pull you" dance is that of cost versus benefit in big government science programs. Do these programs just waste money on politically important public relations campaigns (e.g., beating the USSR in a race to land a man on the Moon)? Or does government-financed "big" science also benefit society in the long run? Based on this small foray into aspects of the Apollo 11 space program, I conclude that indeed, "big" government-funded science does benefit society in unanticipated ways. We now have better fire protection wear, thermal blankets and clothing for recreational and emergency use, Mylar party balloons, MRI machines that use very cold liquid nitrogen and helium to noninvasively image our insides, and specialized, high-efficiency scuba equipment. These are just the "spinoffs" from a few of the technological innovations produced by the space program. We can conclude with Buzz Aldrin's comment to remember another "hidden" benefit of the Apollo Moon missions—a renewed interest in science and how it can benefit society:

> The biggest benefit of Apollo was the inspiration it gave to a growing generation to get into science and aerospace.
>
> Buzz Aldrin

Works Cited

Aldrin, Buzz. Most Famous Apollo Space Program Quotes. https://spacequotations.com/quotes-about-apollo/ (accessed 20190723).

Armstrong, Neil. "Quotes." www.neilarmstrong.com/quotes (accessed 20190825).

———. "Letter to Robert Krulwich." 2010. https://en.wikiquote.org/wiki/Neil_Armstrong (accessed 20190723).

Bryan, Cardis, and Strasburger, William. "Lunar Module Structures Handout LM-5, In support of LM-5 Structures Course," MSC Houston, May 1969, p. 15. https://www.hq.nasa.gov/alsj/a11/a11LM5structures.pdf (accessed 20220614).

"Buzz Aldrin on the Moon in 1969." https://images-assets.nasa.gov/image/as11-40-5903/as11-40-5903~orig.jpg (accessed 20200130).

Cernan, Gene. 1973. BrainyQuotes, https://www.brainyquote.com/quotes/gene_cernan_598282 (accessed 20190613).

Chaikin, Andrew. "Neil Armstrong's spacesuit was made by a bra manufacturer." *Smithsonian Magazine*, Nov. 2013, https://www.smithsonianmag.com/history/neil-armstrongs-spacesuit-was-made-by-a-bra-manufacturer-3652414 (accessed 20190301).

Foster, Lauren. In "Square Astronaut, Round Hole: Chris Hadfield on Risk, Competence, and Leadership," Enterprising Investor, CFA Institute, November 6, 2013. https://blogs.cfainstitute.org/investor/2013/11/06/square-astronaut-round-hole-chris-hadfield-on-risk-competence-and-leadership/ (accessed 20220614).

Garber, Steve. "Challenger STS 51-L Accident January 28, 1986." NASA History, NASA, February 26, 2018. https://history.nasa.gov/sts51l.html (accessed 20190601).

The Graduate. Embassy Pictures, United Artists, 1967.

GYassineMrabetTalk. "Nylon-6,6." This structure was created with PyMOL.—Own work, CC BY 3.0, https://commons.wikimedia.org/w/index.php?curid=9106813 (accessed 20220613).

Jones, Eric M. "PLSS (Portable Life Support System)." Apollo Lunar Surface Journal, 1996. https://www.hq.nasa.gov/alsj/plss.html (accessed 20220130).

Jynto. "Polyethylene Terephthalate." This image was created with Discovery Studio Visualizer, CC0, https://commons.wikimedia.org/w/index.php?curid=15949374 (accessed 20220613).

Kennedy, John F. "JFK Rice Moon Speech." NASA, September 12, 1962, https://er.jsc.nasa.gov/seh/ricetalk.htm (accessed 20220218).

Knight, Laurence. "A brief history of plastic, natural and synthetic." *BBC News Magazine*, May 17, 2014. https://www.bbc.com/news/magazine-27442625 (accessed 20220608).

Kurtus, Ron. "Gravitational Escape Velocity with Saturn V Rocket." First Revision, School for Champions, February 15, 2016. https://www.school-for-champions.com/science/gravitation_escape_velocity_saturn_v.htm#.XWG2w-hKg2w (accessed 20220608).

Mills, Ben. "Teflon." Own work, Public Domain, December 27, 2011. https://commons.wikimedia.org/w/index.php?curid=17821816 (accessed 20220614).

Moore, Jeffrey. "Unit 4. Gas Laws. (PDF)" Moore Chemistry, University of Illinois Urbana-Champaign, 2014, p. 5. http://moorechemistry.weebly.com/uploads/8/9/4/4/8944539/gases_2014.pdf (accessed 20220613).

"Neil Armstrong's Space Suit (EMU)." https://history.nasa.gov/alsj/a11/ap11-S69-38892HR.jpg (accessed 20200130).

Parker, James F., Jr., ed. "Section VI. Systems." SP-368 Biomedical Results of Apollo. Jones, Walton M., ed., NASA 1973. https://history.nasa.gov/SP-368/contents.htm (accessed 20220608).

Pendle, George. *Strange Angel: The Otherworldly Life of a Rocket Scientist*. Mariner Books, 2006.

Shepherd, J.E. "Flash Point and Chemical Composition of Aviation Kerosene (JetA)." (PDF) Nuyt, C.D.; Lee, J.J., Graduate Aeronautical Laboratories, California Institute of Technology, March 2, 2000. https://authors.library.caltech.edu/25832/1/galcit_fm99-4.pdf (accessed 20220304).

Tate, Karl. "NASA's Mighty Saturn V Moon Rocket Explained (Infographic). Space.com, November 19, 2012. www.space.com/18422-apollo-saturn-v-moon-rocket-nasa-infographic.html (accessed 20220608).

Vogt, Gregory L. "Suited for Spacewalking." George, Jane A., ed. NASA JSC Education Working Group, 1998, p. 18. https://www.nasa.gov/pdf/143159main_Suited_for_Spacewalking.pdf (accessed 20220130).

"What is BoPET?" BoPET Films Europe, https://www.bopetfilmseurope.com/what-is-bopet/ (accessed 20220613).

Wolfinger, Kirk, dir. "Apollo's Daring Mission." *NOVA*, DeNoyer, Rushmore; Norton, Sue; Younge, Patrick; Shepheard, Ryan, WGBH, December 26, 2018, Boston, MA, USA. https://www.pbs.org/wgbh/nova/video/apollos-daring-mission/.

3

Selling Space Travel

The Disney Version

CHRIS ROBINSON

For Kate and the great Leonard Maltin

While much of this volume focuses on the cultural implications of the Moon landing, this chapter does somewhat the opposite: it attempts to explore how popular discourse affected the push toward space in the first place, and how it led to the Apollo program in the 1960s. Legendary military and NASA "rocket scientist" Wernher von Braun, the primary architect of both the V-2 rocket from World War II and the Saturn V rocket that propelled astronauts to the Moon, had as his personal crusade the selling of space travel to the public and making it seem viable. In some real sense, the Apollo effort itself rose out of being pre-sold in the popular imagination by von Braun's relentless evangelizing for spaceflight. The space program did not simply arise in a panic after the Russian launch of Sputnik and Yuri Gagarin's first flight. Rather, it was something that had very much been marketed and sold.

Most central to these efforts was Walt Disney's three-part series on "Man in Space," hosted by von Braun, on the *Disneyland* show aired on ABC-TV between 1955 and 1957. For both Disney and von Braun, their work on the series was mutually beneficial—Disney received concepts, ideas, and promotion for his theme park, as well as stellar television ratings—while von Braun received publicity to bolster the case for his particular vision of space travel (see fig 3.1). The show is not unknown and has garnered more attention in recent years; Disney historian Todd Pierce considers it an "attempt to change the domestic policy of the United States from within a

Figure 3.1. Walt Disney and Wernher von Braun.

small soundstage at an animation studio" (203). However, the effect of this television show on the public discourse and what effect, if any, it had on national policy, needs to be more deeply considered.

As a teenager in Germany in the 1920s, von Braun had been enthralled by an article on space travel in a German astronomical magazine, recalling, "Interplanetary travel! Here was a task worth dedicating one's life to!" The experiments of his youthful rocket group, the spaceflight society Verein für Raumschiffahr, soon attracted the attention of the German military, and von Braun had no qualms about persuading his colleagues to work for the Army with him. As von Braun later recalled in a *New Yorker* article, "We felt no moral scruples about the possible future use of our brainchild. We were interested solely in exploring outer space. It was simply a question with us of how the golden cow could be milked most successfully" (Neufeld, *Von Braun: Dreamer of Space* 4). This Faustian bargain, of course, would lead to the V-2 rocket for Hitler and the missile attacks on Great Britain. Charges of opportunism for expediency's sake, with its lack of moral qualms, would dog von Braun for most of his life, though the more serious stories of his use of slave labor did not surface until the 1970s. As the Tom Lehrer song about von Braun went:

> Don't say that he's hypocritical
> Say rather that he's apolitical
> "Once the rockets are up, who cares where they come down?
> That's not my department!" says Wernher von Braun
>
> Some have harsh words for this man of renown
> But some think our attitude
> Should be one of gratitude
> Like the widows and cripples in old London town
> Who owe their large pensions to Wernher von Braun (Lehrer)

Even von Braun's plan at the end of World War II to make sure he ended up on the American side was colored by this opportunism. Von Braun's brother had made a trip to America in 1935, and glowingly told him that "America is the place to build your Moon rockets." In a secret conversation with an engineer at the end of 1944, von Braun revealed his plans if the war was lost, saying, "I will offer my services to the Americans . . . and then I will build my space rocket" (Neufeld, *Von Braun: Dreamer of Space* 290).

For von Braun, the vision of manned spaceflight was the main thing—in fact, really, the only thing, superseding all else. This is what space historian Dwayne Day in 1994 termed the "von Braun paradigm," with large, expensive, manned spaceflights to far-off destinations being favored over cheaper, unmanned, robotic missions, as were favored by other space scientists such as James Van Allen. As Michael Neufeld writes, von Braun "was simply uninterested in any other form of exploration. For him, as for other space advocates of his generation, sending humans was the point" (Neufeld, "The Von Braun Paradigm" 331). "Von Braun's disinterest in robotic probes and his obsession with monumental human exploration," as Neufeld puts it, was in some sense a practical marketing scheme in terms of what might pique the interest of politicians and find favor with the public.

In a later interview, von Braun qualified his ambitions, saying that he had "always felt that it would be a good idea to read the signs of the times and respond to what the country really wants, rather than trying to cram a bill of goodies down somebody's throat for which the time is not ripe or ready" (Neufeld, "The Von Braun Paradigm" 335, 340). The public first had to be sold on the idea, and von Braun was more than willing to prime the pump, as it were.

Neufeld's biography of von Braun takes on a more nuanced approach. Despite his flights of fancy in the 1950s, von Braun's rocket designs, such as the Saturn V, were fairly conservative. The Redstone/Loki setup he cobbled together for the first Explorer satellite rocket designs was, frankly, cheap and somewhat improvised. As Neufeld notes, for Explorer, "von Braun's launcher idea was the exact opposite" of his grander schemes—"it was to be simple, low-budget, and technologically conservative, highlighting the apparent contradictions of his character." To Neufeld, "von Braun did not feel this contrast as a contradiction; he distinguished between short-term and long-term perspectives and believed that in order to sell space to the public, he had to go for the grand scale" (Neufeld 282–283).

After the war, von Braun and his team of scientists and engineers were brought over from Germany in 1945 and 1946 under the US government's Operation Paperclip program. Once ensconced in Fort Bliss, Texas, they were put to work launching V-2s for the Army's missile program. Conditions were Spartan at first; the engineers (who referred to themselves as Prisoners of Peace) were initially only allowed off the base grounds (with escort) once a month (Stuhlinger 82–85; Neufeld, *Von Braun: Dreamer of Space* 82).

Washington's plans for the German team were only concerned with military weapons, and budget cuts in 1947 under the Truman administration led to belt-tightening. As Neufeld writes, faced with "a lack of popular support, von Braun would turn to the idea of selling the American public on spaceflight. Since it obviously was not going to come about simply as a by-product of military development, he had to convince ordinary people that space travel was not a silly or utopian proposition" (Neufeld, *Von Braun: Dreamer of Space* 224). As fellow German engineer Ernst Stuhlinger puts it, von Braun realized that in order to pursue his dreams of space travel, "he would have to achieve a number of other goals first, the most important being that, as von Braun would say, 'the idea of spaceflight must be popularized to arouse public enthusiasm'" (Stuhlinger 118). This was a time when, as Harlen Makemson puts it, von Braun "pondered how to project his vision onto the consciousness of the American public" (Makemson 5).

The first attempt in this vein was von Braun's ill-fated *The Mars Project*, a science-fiction novel that he wrote with the aim of popularizing spaceflight. Von Braun seized upon the idea in 1947 and wrote the novel in 1948 and 1949 while in Texas. It involved a meticulous manned mission to Mars consisting of no less than ten ships; the book being planned out with the assistance of half a dozen of his fellow Germans in Texas. Written in German and then translated to English, the manuscript wound up being rife with technical details; the scientific appendix attached to the book alone ran to some 120 pages. When von Braun met with publishers, they found the book lacking in drama. An editor with Doubleday remarked, "'Mars Project' is a fascinating book—but not, I'm afraid, from the standpoint of a potentially saleable novel . . . the only thing that I can think of that we might do with it is build a rocket ship." Later, von Braun would lament the reactions of the publishers to a fellow space enthusiast, saying the reactions ran along the lines of: "Who's interested in the Mars atmosphere or the initial thrust of a satellite ship? The story lacks a girl!" (Neufeld, *Von Braun: Dreamer of Space* 246). In the end, the book was rejected by over a dozen publishers, though the technical appendix ultimately wound up being published separately in Germany—and years later, in the United States.

Von Braun's next major project to popularize space wound up being far more successful. By mid-1950, his group had moved from Texas to the Redstone Arsenal in Alabama, later to become NASA's Marshall Space Flight Center. Von Braun began making the rounds of various astronautical conferences to pitch his ideas, sometimes delivering his ideas in person, and sometimes sending a surrogate to deliver a paper when international travel proved too difficult. At a conference in San Antonio in 1951, von Braun ran into *Collier's* magazine editor Cornelius Ryan (later better known for his World War II books such as *The Longest Day* and *A Bridge Too Far*) at a bar. *Collier's*, which ultimately folded in 1957, was a mass-market magazine experiencing declining circulation, and Ryan's superiors had asked him to attend the conference, wondering if there might be any potential in an article on spaceflight. Over a long evening of alcohol, von Braun and two other space enthusiasts, including the chairman of the Harvard astronomy department, worked hard on convincing the editor of the validity of spaceflight and finally by the end of the night had made him a believer.

Figure 3.2. This issue of *Collier's* from March 1952 contributed to the American public's fascination with the possibilities of space.

This led to the famous *Collier's* articles on space, which began with the cover story, "Man Will Conquer Space Soon," in March 1952 (see fig 3.2)—for which von Braun recruited numerous specialists, including *Conquest of Space* author Willy Ley and space artists such as Chesley Bonestell, also known for his work on George Pal's *Destination Moon*. There were numerous sections and articles besides von Braun's feature piece, including an editorial asking, "What are we waiting for?" Von Braun was flown to New York for a huge publicity tour by *Collier's*, including appearances on the *NBC Nightly News* and the *Today* show, among others. On his return to Huntsville, Alabama, *Collier's* publicist Seth Moseley sent him a congratulatory note, lauding his "magnificent cooperation" and saying, "you are as successful a salesman as you are a scientist" (Neufeld, *Von Braun: Dreamer of Space* 260). The issue generated quite a stir and led to increased discussion of the possibility of exploring space. Some certainly regarded von Braun as a crackpot or a loon, but others did not.

Over the next two years, following on the success of that first issue, von Braun and his collaborators would put out no fewer than seven additional *Collier's* sections exploring the possibilities of space, including "Man on the Moon" (see fig 3.3), "The Baby Space Station," "Can We Get to Mars?," and others, sometimes accompanied by additional publicity blitzes. The impact of the *Collier's* series is difficult to accurately quantify, but it seems to have been considerable. David Meerman Scott and Richard Jurek note that a Gallup poll in 1949 found that 15 percent of the American public thought it possible that "men in rockets will be able to reach the [M]oon in the next 50 years." By 1955, that figure had moved to 38 percent. "The *Collier's* series," they note, "is believed to have been a primary factor in this increase" (Scott and Jurek 9). Ron Miller similarly noted, "it is difficult to overstate the effect of the articles." (Makemson 3) A further push would come in 1954 when von Braun would enter into his collaboration with Walt Disney.

By 1954, Disney's studio had largely recovered from its wartime setbacks, the damage from its animators' strike and postwar austerity, and was able to make full-length feature animated films again rather than anthologies. Disney had also embarked on what was to be his most ambitious venture of the decade, the theme park known as Disneyland.

The original idea for Disneyland seems to have sprung from a trip Disney had made to the Chicago Railroad Fair with animator Ward Kimball. Kimball was one of the more eccentric of Disney's veteran animators; besides being a jazz musician, he ran a full-sized (not model) steam train in his backyard. Kimball assumed that Disney thought he would be a good traveling companion to the fair. Kimball remembers Disney being with him on the numerous train rides, factory displays, and nostalgic recreations of villages: "we were like two kids," Kimball recalled. Following the trip, Disney began having his artists draw up plans for a small-scale amusement park on the Disney studio lot, complete with a model train and a period small town. While this park was never built, it did serve as the genesis for Disneyland, which was ultimately built on 160 acres of orange groves in Anaheim, California.

In 1953, Disney entered into an agreement with the ABC television network to develop the *Disneyland* TV series. The money from the series and loans from the network would help to finance construction of the theme park, while the show itself would serve as a weekly hour-long advertisement for the park. The show would be organized around the four initial theme-park areas: Adventureland, Frontierland, Fantasyland, and

Figure 3.3. One of seven additional *Collier's* features exploring the promise of space travel.

Tomorrowland. Provisioning content for the first three areas was not an issue. Disney had lots of animated films with princesses that could be used for Fantasyland content, its successful *True-Life Adventure* documentary films could be repurposed for Adventureland, and the Fess Parker *Davy Crockett* series would be filmed for Frontierland.

Tomorrowland, however, was an issue; there had been no previous Disney work involving the future. In March 1954, Disney proposed the idea of having Kimball head a unit devoted to the Tomorrowland segments. Besides being a natural fit for Kimball's technological interests, it would give him something to do, as Kimball's more modernistic style of animation, as in his *Toot, Whistle, Plunk and Boom* (1953), was no longer a good match with much of Disney's feature animation. An early idea on developing a show about atomic energy fell apart, and Disney—overworked with construction of the park—was looking for ideas. According to one of Kimball's accounts, "Walt came to me and said, 'You guys are the modern thinkers around here—can you think of anything we can do on Tomorrowland?'"

(Smith 55). Another version has Disney asking Kimball to develop something for Tomorrowland, saying "You're interested in UFOs and all that stuff" (Pierce 188). According to Kimball, "That's when I said I had been following some very interesting articles on space in *Collier's* magazine" (Smith 55). Kimball lent Disney three issues of *Collier's* and the agreement was made to move forward on a treatment. By mid-April, Disney was impressed with the work and gave Kimball the approval to move forward. Disney mimed giving Kimball a blank check in one meeting, telling Kimball, "Write your own ticket" (Stuhlinger 115). With Disney intensely preoccupied with Disneyland at the time, Kimball would have relatively free rein creatively. Ultimately, there would be three space-related episodes, in large part imitating the *Collier's* articles. The first, "Man in Space," would deal with the launch of a spaceship, the second with a trip around the Moon, and the third with a trip to Mars.

For Kimball, "it was fascinating to me to realize that there were these reputable scientists who actually believed we were going out into space" (Stuhlinger 115). Kimball reached out to Willy Ley, who had worked on the *Collier's* articles, and by April, Ley was on the Disney lot. At Ley's urging, other collaborators from the *Collier's* articles were brought on, and von Braun was by some accounts at the top of Kimball's list.

Von Braun at the time was involved in another venture, trying to develop a space show called *Target for Tomorrow* for CBS; but once that project failed to come to fruition, he swiftly moved to sign on with Disney around June 1954. He would serve as a technical advisor, help Disney develop models for the show, appear on camera as an expert, and participate in the writing of the script. The designs from *Collier's* would need to be tweaked slightly to avoid copyright infringement; thus, von Braun's space rocket from the first show went from three stages to four (see fig 3.4).

Von Braun apparently made his first trip to the Disney studios in July 1954, when he supposedly had his first meeting with Disney. By one account, Disney was left open-mouthed by von Braun's spiel. According to Kimball, when Disney "started talking to von Braun, he got really enthusiastic." In a later meeting in Kimball's office, Kimball recalls that von Braun read an article tacked to Kimball's wall about Russian developments for possible space exploration. As Kimball recalls, "von Braun stood there and read the whole thing; and then he turned to me and said, 'Well, Ward, we'd better get busy.'" As Todd Pierce writes, for von Braun, the Disney project in part "served to build taxpayer support so that Americans might beat the

Figure 3.4. Ward Kimball and Disney artists building spaceship models for "Man in Space." Valley Times Photo Collection, Los Angeles Public Library.

Russians into space. The absurdity of it all was this: that some of the best scientists and engineers in the country would need to make their pitch through a company that primarily produced family and children's entertainment" (Pierce 197–198).

Von Braun, for his part, worked on the script from Huntsville, but he managed to squeeze in numerous trips to Disney over the summer and fall for technical meetings and story conferences, often working late into the night with Kimball's team on details such as fuel lines. The intent was to make the show as technically accurate as possible. A memo from Kimball's research assistant, Bill Bosche, in August 1954 says to von Braun, "we mailed to your home in Alabama yesterday the drawings on the space wheel, spacesuit, and the three-stage instrument carrier rocket . . . One further question—was the exact altitude reached by the V-2 100 or 114 miles?" (Pierce 200). Here, for von Braun, were finally people who were interested in the details.

Live-action shooting with von Braun and other space experts was done in the fall of 1954, and work continued on the other episodes. For greater believability (and less cost), the Moon trip would be live-action rather than animated. The Moon episode did not have a Moon landing, but only a ship orbiting the Moon. Ernst Stuhlinger speculated that this was due to Disney's desire for what he had termed "a factual science presentation" (Stuhlinger 116). The Mars show could be more speculative. In any case, the second show would be largely von Braun's; he would be the only on-camera "expert" appearing in the show. Von Braun paid meticulous attention to the technical details, supervising the Moon ship, and at one point even adding a tint to the glass of the spacecraft. In a move indicative of this obsession with detail, Kimball even hired von Braun under a separate contract to actually develop and work out a real, accurate, viable flight path to the Moon. Von Braun spent several months working on the calculations in the fall of 1954. A letter from Kimball associate Bill Bosche to von Braun in January 1955, following up on the flight-path calculations, shows how deep Kimball's team was into the weeds on this project as compared to a typical Disney cartoon. Asked Bosche: "Would check points and time intervals and velocities be the same on the return to earth as the [M]oon-bound portion of the ellipse?" (Pierce 207). The show would not just look like a doable idea for sending a rocket to orbit the Moon; it would demonstrate the real thing. This befuddled one of the artists, who at one point interrupted a spirited conversation between von Braun and Ley on how the tracking instruments at ground control would work with the observation: "You know, here at Disney's we just *draw* things" (Pierce 207).

By January 1955, a story was widely circulated by the United Press claiming, "Space Satellites Possible in 1957, Scientist Says." At the meeting of the National Capital Astronomers, Prof. Fred Singer, of the University of Maryland, said that the time to start spaceflight was "right now"; that rockets "could be built with present engineering knowledge"; and that starting with a satellite "would pave the way for flight into interplanetary space." The article was widely quoted in the months to come ("Space Satellites" 57).

By March 1955, the first *Disneyland* episode, "Man in Space," was ready to air (fig 3.4). Kimball sent a postcard to von Braun, reminding him to "Tell the boys at the plant that Mr. Mickey Mouse of V-2 fame will hold forth," to which von Braun had replied to Bosche, "Brother, this is the only topic in our home!" (Neufeld, *Von Braun: Dreamer of Space* 291). The show opened with Walt Disney in his office, saying, "Great new discoveries have brought us to the threshold of a new frontier—the frontier of interplanetary

Figure 3.5. Walt Disney promoting space travel. © Walt Disney Productions

space" (see fig 3.5). Von Braun, in his on-camera segment, made a prediction that would seem bold at the time: "If we were to start today on an organized and well-supported space program, I believe a practical passenger rocket can be built and tested within ten years" (Smith 54, 58).

When "Man in Space" aired on March 7, the ratings were unexpectedly strong, even in those days of only three major networks, and the *Disneyland* show climbed that year into the top ten most popular shows of the season (Jones 57). Von Braun "costar" Willy Ley noted that the viewership for the show was "higher than we thought it would be" (Pierce 208). An article in the San Pedro *News-Pilot* noted that von Braun and the scientists "say that we're not too far from some kind of interplanetary travel," and that "the heady tonic of Sunday supplement stories" was "turned, in our day, to something breathtakingly close to science fact" (Freeman 26). An article in the *Los Angeles Herald* by Scholer Bangs was more effusive, proclaiming,

> Walt Disney may be America's "secret weapon" for the conquest of space! Apparently, and quite by accident, he has discovered the "trigger" that may blast loose his country's financial resources and place the Stars and Stripes of the United States aboard the first inhabited Earth satellite . . . it is so near that we can practically feel the Earth tremble under the rocket blast of Dr. von Braun's spaceship. "Man in Space" is believable, and Disney has close to 100,000,000 Americans

> believing. Half of the voting population of the USA has probably reached two impressive conclusions; "It CAN be done!" and "Let's get on with it!" (Smith 59)

A repeat broadcast of the show was scheduled for June 1955. Far more so than the *Collier's* articles, the Disney space shows provided a great boost to popularizing the idea of a government space program. As Disney historian David R. Smith notes, von Braun was well aware of the reach of television: "When Disney called him to do the space shows for TV, he pounced on the opportunity. He realized the power of TV, with millions of people watching. You just don't get those numbers reading *Collier's*" (57).

April and May 1955 saw the release of George Pal's *The Conquest of Space*, based on the Ley and Bonestell book, focusing on a manned voyage to Mars. The film itself was not a success; with its creaky script, it was not one of Pal's best efforts, and it marked the end of Paramount's ventures into larger-budget science fiction. However, the discourse surrounding the picture illustrated the growing awareness of space in the public consciousness. A syndicated article on the film in the Owensboro *Messenger-Inquirer* noted that "public interest is currently so high in the possibility of man's occupation of other planets," and that the film "goes beyond many of the headlines in today's news" ("'Conquest of Space' Is New Exciting Adventure Thriller" 27). An advertisement for the film in the *Charlotte News*, besides offering free spacesuits to the first few children in line, promised that a Mars landing "will happen within your lifetime!" and quoted from *Collier's* and the Singer article about satellites being possible by 1957 ("Conquest of Space" newspaper advertisement, *The Charlotte News* 11). Another advertisement for the film in the Kokomo *Tribune* also quotes the Singer article and speaks of "the amazing new frontier which will be fought for and won during our lifetime" (fig 3.7). The future conflict with Soviet Russia is presaged by a banner reading: "Race into space—Reds want first space platform." A timeline at the bottom of the advertisement highlights scientific achievements (corrected here): "[1945]—the atom bomb, [1950]—the hydrogen bomb, 1955—the atomic submarine, 1960—mankind's greatest adventure." We still have not had a manned mission to Mars, but this ad is eerily prophetic; few in 1955 would have predicted that we would have men in space by 1961 ("Conquest of Space" newspaper advertisement, *The Kokomo Tribune* 14).

While *Conquest of Space* was ostensibly based on the Ley and Bonestell book, much of the film's technical design, including the space station, came

Figure 3.6. Newspaper ad from the New York *Daily News*, March 1955.

from von Braun's work and the *Collier's* article, leading von Braun to be listed on the Internet Movie Database as an (uncredited) technical adviser and given credit for the story. While director Byron Haskin recalled von Braun being on the set as a technical adviser, there is no mention of this collaboration in any of the voluminous biographies of von Braun (Haskin 230).

Figure 3.7. Newspaper ad for *Conquest of Space*, April 1955.

When "Man in Space" repeated on ABC in June 1955, it wound up having a viewer of particular importance—President Dwight D. Eisenhower. According to Ward Kimball, following the repeat, Eisenhower "borrowed the show to run for the brass in the Pentagon. He called Walt personally, and they ran it for a couple of weeks" (Smith 59). In Kimball's retelling, Eisenhower had been "absorbed" by the program but "realized that he had generals in the Pentagon who didn't understand or accept these ideas." At first, the military didn't support America's program for space exploration. By Kimball's account Eisenhower "flew them all in, and for two weeks, he screened our program for his top generals" (Pierce 211). Historian Mike Wright has called this account "difficult to verify" (8). As Kimball biographer Todd Pierce writes, however, Kimball's story may only be "partially accurate," but still it "presents an amazing picture: a small group of artists at a family animation studio had produced a TV show that was being used by the president to help determine policy" (Pierce 211). The Russians, for their part, verifiably also asked for a print of the program from Disney, which was refused. While Disney was a known anti-communist, according to Kimball, the real reason was resentment toward the Russian government for not returning a loaned print of *Snow White* for over a decade, finally sending back a work print with Russian subtitles that had been screened countless times (Scott and Jurek 12–15).

Work continued through July on the second show, "Man and the Moon," for which von Braun would be the primary star and sole on-camera expert. Von Braun's schedule shows three days of live-action shooting from the 11th to the 13th, squeezed in between meetings with contractors, presentations of satellite ideas to military brass, and meeting with the Stewart committee, the high-level commission set up to choose among possible satellite projects (Neufeld, "The Von Braun Paradigm" 295).

July 17, 1955, finally saw the opening of Disneyland, along with its Tomorrowland and its von Braun–inspired trip to the Moon. After a rough opening day, with overcrowding due to counterfeit tickets and a slew of breakdowns, the park nevertheless rapidly became a great success for the Disney organization, only taking two months to welcome its one-millionth visitor. In the Wilmington *Morning News*, Fred Othman called Disneyland "eye-popping" and a "fabulous financial success," raving about "what the incredible Disney calls Tomorrowland," with its TWA Moonliner rocket "flying regular round-trip schedules to the moon" (22).

The full extent of von Braun's actual involvement in the design of the Moonliner ride and the rest of Tomorrowland, including Space Station X-1, varies in the von Braun histories, though it is clear that von Braun was to some degree a participant. Michael Neufeld's biography of von Braun says that "beyond glancing at the plans, Von Braun had little impact on the ride" (Neufeld, *Von Braun: Dreamer of Space* 288). By comparison, Catherine Newell says the ship "was personally designed for Disneyland by von Braun and Ley" (423), and Bob Ward's biography says that von Braun was paid for consulting work "on Tomorrowland at the original Disneyland, then on Walt Disney World" (103). At the very least, it seems safe to say, as Scott and Jurek do, that the TWA Moonliner (an early example of corporate branding) was at least "designed by Disney's John Hench with advice from von Braun." Plastic model kits for the Moonliner were widely sold, along with models of ships from the "Man in Space" films. Ironically, when the Moonliner model was sold in the UK, the British manufacturer removed the Disney name and sold it instead as "Dr. Werner von Braun's Moonliner" (see fig 3.8), a remarkable feat of "personal rebranding" for someone whose V-2s had rained death on England in the previous war (Scott and Jurek 14).

For Newell, the effect of the space aspect of the Tomorrowland park itself, in addition to that of the Disneyland series, cannot be underestimated. "While the cause of space exploration really could not have asked for a better champion than Wernher von Braun . . . the whole enterprise hit the metaphorical jackpot when it was adapted into a television special by Walt Disney—doubly so when it became part of the Disneyland experience" (423). In attempting to understand the motivations behind the push for space, Newell argues, we must look beyond the simple notion of the "space race" with the Russians. In the standard history, "the Space Race was an escalation of innovation triggered by Sputnik in 1957 and an extension of the nuclear arms race between the United States and the USSR." To Newell,

> As an answer to the question of why "space" became a cultural phenomenon—why millions of Americans during the 1950s became obsessed with exploring outer space, with building rockets to the [M]oon, and establishing colonies on Mars, the explanation of "military high ground" falls flat . . . [T]hrough the process of "Disneyization" . . . the dream of space flight became indoctrinated in American

> culture . . . [I]t is only by understanding the content of Tomorrowland that we can finally understand the breathless enthusiasm with which America greeted the space age. (426–427)

On July 29, 1955, the push for space started in earnest, with President Eisenhower's announcement that the United States would undertake the launching of a scientific satellite for the International Geophysical Year of 1957.

Figure 3.8. Moonliner model kit.

The reasons for this initiative were numerous, including the push from the science community to do something for IGY and the knowledge that the Russians were working on their own satellite program. The Soviet Space Commission had been announced on April 16 of that year, with Moscow radio proclaiming weeks later that a satellite was planned, and the expectation was that the Russians shortly would make an official announcement at the IAF conference in Copenhagen (which they did, a mere four days after Eisenhower). There was also evidence of Eisenhower's being persuaded by the concept of "overflight," the idea that a satellite would set the precedent for space being a borderless, international territory, laying the groundwork for photo-reconnaissance and surveillance satellites (Neufeld, *Von Braun: Dreamer of Space* 294). These concepts had followed the rejection of Eisenhower's proposal for "open skies" aircraft photo-reconnaissance earlier in the month. Many, however, viewed the impact of Disney and von Braun's "Man in Space" program as a crucial factor in Eisenhower's decision.

Following Eisenhower's announcement, Disney and ABC made plans for a third airing of the first "Man in Space" episode in September in order to capitalize on the publicity of the new American satellite program. Kimball wrote von Braun on August 24, "the studio intends to ballyhoo the first 'Man in Space' show when it's rerun for the third time on September 7 as the one item that contributed the most to Eisenhower making the space satellite announcement." A concerned von Braun replied on August 30,

> I am sure that a statement to the effect that this show contributed the most to President Eisenhower making the satellite announcement is bound to antagonize quite a few people who had a hand in putting the project across. They will feel that I myself through the vehicle of the Disney Studio am trying to get credit for more than I deserve and this would hurt my pleasant relations with many of the people I have in mind. I would appreciate it if you would take up this matter with Walt and try to have the pitch changed.

On September 1, a chastened Kimball replied, "I am sorry you became so alarmed . . . please file my remarks under the headings of 'Colorful Expressions'" (Smith 59). Von Braun's hesitancy here is understandable, for he had to deal with many factions in his quest for outer space, and had to tread carefully—rightfully so, in view of Eisenhower's later complaints to his aides about von Braun's being "publicity-seeking" (Ward 316). Note, however, that von Braun did not expressly *deny* that the Disney show was partly responsible for Eisenhower's announcement, rather that it would make others who contributed angry and, as he says later in the letter, "would hurt the cause far more than it would help" (Smith 59).

No matter Eisenhower's real reasons, much of the press thought that von Braun's Disney program had been a motivating factor. A syndicated column by John Lester the week after Eisenhower's announcement ambitiously claimed,

> Several congressmen have credited Disney's "Man in Space" films, as shown on ABC-TV's "Disneyland" series, with exerting final pressure on the government to get its multi-billion dollar space project out of mothballs. The project, as recently announced, was designed to eventually land a spaceship on the moon [*sic*], and the mountain of letters and wires received in Washington from space fans, urging that something be done NOW, dated from Disney's first space telecast. (Lester, "Radio and Television" 21)

Eisenhower's announcement similarly increased visibility for Disney and the *Disneyland* series. On August 14, the Quad-City *Times* also remarked that "Since President Eisenhower's approval for launching a man-made satellite to circle the earth [*sic*], increased importance and interest is surrounding 'Man in Space' . . . Great are the similarities between the government's program and Disney's visualization" ("Disney to Exploit Outer Space" 60). A longer version of the same syndicated column ran in Binghamton's *Press and Sun-Bulletin* with the headline, "Ike's OK on Satellite Boon to Disney Show" ("Ike's OK" 30). John Lester again noted that Disney had struck gold with this repeat, "because of interest in space things since the government's 'satellite' announcement" (Lester, "Title Fight Will Be On Radio" 3). Similarly, the Racine *Journal Times* noted that "the Disneyland repeat for tonight takes on special significance because of the recent approval by President Eisenhower of the launching of a man-made satellite" ("Television and Radio Programs" 21). An article in the *Daily World* also noted, "Disney and 'Man in Space' have indirectly been credited with prompting President Eisenhower to disclose this country's plans to launch a man-made satellite." Here was synergy for both von Braun's vision and Disney's ambitions.

Von Braun's hopes to launch the first American satellite were defeated (at least temporarily) in September 1955 when the administration decided to instead move forward with the Navy's ill-fated Vanguard project, which intended to launch the first satellite using the Naval Research Laboratory's

Vanguard rocket rather than with something from von Braun's group. Disney's preparations for the second episode, "Man and the Moon," moved forward, and it was set to air on December 28, 1955. The show generated another wave of publicity for both Disney and von Braun. Larry Wolters, in his *Chicago Tribune* article, "Mr. Disney Shows Way to the Moon," echoed the earlier Bangs article: Disney had pulled the trigger to "blast loose this country's financial resources and place the Stars and Stripes aboard the first space ship." The magazine added that "scientists who participated in the filming of the Disney show are certain that, thru public financial support, a space project will become readily available."

Wolters noted that, following Eisenhower's satellite announcement, Sen. Charles T. Curtis of Nebraska called the attention of his colleagues and the nation to the achievements of Disney. "Few know," he said, "that in addition to his role in the entertainment world, Walt Disney's technical abilities have been of great service to our government." (This speech had taken place four days before Eisenhower's announcement at the end of July.) The article concluded by saying that a double-page color spread of scenes from the film could be seen in the upcoming Sunday magazine, as well as designs of von Braun's space station (14). This was widely syndicated in other newspapers, such as the Cedar Rapids *Gazette* ("'Man and the Moon,' Disneyland Special" 7). A December 30 article in the Louisville *Courier-Journal* by Joseph Landau on von Braun called him a "space travel salesman," and noted that "it was probably von Braun, as much as anybody, who sold the United States on going ahead with a space satellite." However, Landau chided, "to space travel fans, von Braun is something of a prophet. To rocket experts, he is a sound technician who has something of the charlatan in him. They feel that he knows his business when it comes to rockets, but that he's riding his enthusiasm too hard" (Landau 5). Nevertheless, the entire scheme was working.

The completion of the final episode of the planned three, "Mars and Beyond," was put on hold through 1956 and most of 1957, as Kimball's Disney unit, now the acknowledged experts on space films, had been asked by the National Academy of Sciences to produce a film on the Vanguard satellite project. As Kimball associate Bill Bosche recalls, by late 1957, they "were just about to go into production when the Russians put up the Sputnik, and of course that cooled the whole thing." Kimball in a later interview recalled being distraught that months of work were gone, but "There was nothing to do but junk the whole project" (Pierce 214). The next day, Kimball's team arrived to find that Kimball had hung a black wreath on the Vanguard storyboards.

Sputnik's launch made von Braun seem the prophet of the space age, as the *New York Times Magazine* then called him, and by November he was on the cover of *LIFE* magazine, with the article calling him "The Seer of Space." While von Braun appears in the Mars film, his role was smaller since he was now working on launching a real space vehicle, although he had already done considerable work on the story and structure previously. The Vanguard program continued to have issues, and von Braun was allowed to continue working on his ideas to orbit the first satellite with his Redstone rocket as a tacit backup, albeit with numerous obfuscations. Many test launches were carried on under the guise of nose cone research, "reentry test vehicles," etc.; and on one occasion, the military was actually sent to oversee a test launch to make sure that von Braun did not "accidentally" launch a satellite.

"Mars and Beyond" aired on December 4, 1957, and received stellar ratings and excellent reviews; *Time* magazine called it "Disney's prize effort of the season," while one pundit commented that its timeliness was especially propitious since this was a time "when man's achievements in space are beginning to catch up with his thoughts and dreams" (Smith 62). (See fig 3.9.) Two days later, the Vanguard rocket spectacularly blew up on the launching pad on live television, with Walter Cronkite watching. This failure was famously derided as the "Flopnik," and gave von Braun another opportunity. Despite numerous follow-up attempts by the Vanguard project to make a proper launch, its problems persisted—and instead, von Braun's Redstone rocket successfully launched the first American satellite, Explorer 1, on January 31, 1958. This landed him on the cover of *Time* magazine in February.

Von Braun's highest ambitions had always actually been for a manned Mars landing, as explained in the Disney film and his numerous works outlining a possible trip to Mars. He had laid out plans for a Moon mission in 1957, and later the next year he was among those proposing a national space agency, which later became NASA. His first love was always Mars, but he was pragmatic enough to be, as Neufeld describes it, in favor of a "shortcut" to a manned lunar landing (Neufeld, *Von Braun: Dreamer of Space* 356). It was cheaper and ambitious enough to give the Americans time to catch up to the Russians, but doable enough to be completed by the end of the decade. Besides, as von Braun remarked when discussing his proposals made

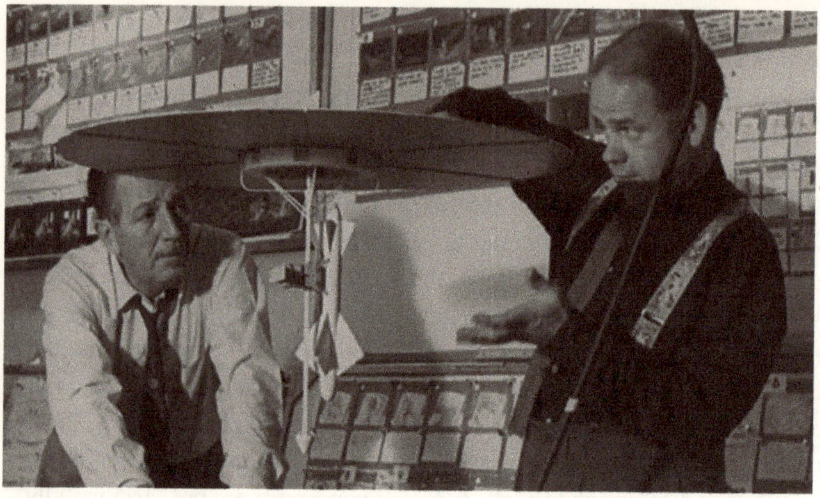

Figure 3.9. Ward Kimball and Walt Disney going over models for "Mars and Beyond." © Walt Disney Productions.

to the later Kennedy administration, "everyone knows what the Moon is" (Ward 128).

The ramifications of the Moon landing and the Apollo project have been well covered in the other chapters here, and the history of the Apollo missions certainly needs no recap. For von Braun, though, the ultimate goal was still Mars, as in the culmination of his Disney films. Following the landing of Apollo 11, on August 4, 1969, von Braun presented a detailed proposal for a manned landing on Mars in 1982, only to be rejected by the Nixon administration, with its desire for cost-cutting after the ruinous expense of the Vietnam War. Unlike the Disney programs, there would be no final episode climaxing in a manned Mars landing.

Disney and von Braun stayed in contact over the years, with von Braun inviting Disney and his team to tour the manned spaceflight center in Huntsville in 1965 (fig 3.10). According to von Braun's papers, he had hoped to sell Disney on another film about manned spaceflight, though nothing ever came of it (Scott and Jurek 15).

Animator Ward Kimball's last contact with von Braun came in December 1968, when Apollo 8 became the first manned mission to orbit the Moon. In truth, this was not the original plan for Apollo 8, which had involved testing the lunar module in low Earth orbit. That plan was threatened by delays and defects in the lunar module; ultimately, it became the

Figure 3.10. Disney (*right foreground*) with von Braun (*center*) at the Marshall Space Flight Center.

mission of Apollo 9. The idea to send Apollo 8 all the way to the Moon (sans lunar module) was the brainchild of George Low, the manager of the Apollo Spacecraft Program Office, who brought the proposal to von Braun on August 9, 1968. According to the meeting minutes, von Braun was enthused and said something to the effect of "Why not? Once you fired up the Saturn V, it did not matter how far it went" (Neufeld, *Von Braun: Dreamer of Space* 424). On December 24, 1968, Apollo 8 entered lunar orbit and circled the Moon just as von Braun's rocket did in the Disney show from

1955; its flight plan similar to the one von Braun had designed for Kimball. That night, Kimball received a surprise call from von Braun, who told the animator, "Well Ward, it looks like they're following our script" (Pierce 241). In some strange sense, the American space program had—at least in part—literally come out of a Disney TV show.

Works Cited

"Conquest of Space" newspaper advertisement. *The Charlotte News*, 20 June 1955, p. 11.

———. *The Kokomo Tribune*, 23 Apr. 1955, p. 14.

"'Conquest of Space' Is New Exciting Adventure Thriller." *Owensboro Messenger and Inquirer*, 29 May 1955, p. 27.

"Disney Sets 'Man in Space.'" *Daily World*, 4 Sept. 1955, p. 24.

"Disney to Exploit Outer Space on TV." *Quad-City Times*, 14 Aug. 1955, p. 60.

Freeman, Donald. "Disneyland Epic Took Year to Make." *San Pedro News-Pilot*, 18 Mar. 1955, p. 5.

Haskin, Byron. *Byron Haskin: An Interview by Joe Adamson*. Metuchen, New Jersey, The Directors Guild of America and Scarecrow Press, 1984.

"Ike's OK on Satellite Boon to Disney Show." *Press and Sun-Bulletin*, 28 Aug. 1955, p. 30.

Jones, Will. "A 51-Acre Beehive." *Minneapolis Sunday Tribune*, 19 June 1955, p. 57.

Landau, Joseph. "Space-Travel Salesman: Rocket Authority Von Braun Has His Eye Fixed on Mars." *The Courier-Journal*, 30 Dec. 1955, p. 5.

Lehrer, Tom. *Songs and Lyrics by Tom Lehrer*, https://tomlehrersongs.com/.

Lester, John. "Radio and Television." *York Daily Record*, 4 Aug. 1955, p. 21.

———. "Title Fight Will Be on Radio, Rooney Will Do 'Live' TV." *The Charlotte News*, 15 Aug. 1955, p. 3.

Makemson, Harlen. *Media, NASA, and America's Quest for the Moon*. Peter Lang Publishing, 2009.

"'Man and the Moon,' Disneyland Special, At 6:30, Channel 9." *The Gazette*, 28 Dec. 1955, p. 7.

Neufeld, Michael J. *Von Braun: Dreamer of Space, Engineer of War*. Vintage Books, 2007.

———. "The Von Braun Paradigm and NASA's Long-Term Planning for Human Spaceflight" in *NASA's First 50 Years: Historical Perspectives*, edited by Steven J. Dick. NASA, 2010, pp. 325–347.

Newell, Catherine L. "The Strange Case of Dr. Von Braun and Mr. Disney: Frontierland, Tomorrowland, and America's Final Frontier." *Journal of Religion and Popular Culture*, vol. 25, no. 23, 2013, pp. 416–429.

Othman, Fred. "Fred Othman Reports: Disneyland Draws 20,000 Visitors a Day." *Wilmington Morning News*, 12 Aug. 1955, p. 22.

Pierce, Todd James. *The Life and Times of Ward Kimball: Maverick of Disney Animation*. UP of Mississippi, 2019.

Ryan, Cornelius, et al. "Man will conquer space SOON!" *Collier's Weekly Magazine*, 22 Mar. 1952.

Scott, David Meerman, and Richard Jurek. *Marketing the Moon: The Selling of the Apollo Lunar Program.* MIT Press, 2014.

Smith, David R. 1978. "They're Following Our Script: Walt Disney's Trip to Tomorrowland" *Future,* vol. 1, no. 2, 1978, pp. 54–63.

"Space Satellites Possible by 1957, Scientist Says." *The Boston Globe,* 9 Jan. 1955, p. 57.

Stuhlinger, Ernst, and Frederick Ordway III. *Wernher von Braun, Crusader for Space.* Krieger Publishing Company, 1996.

"Television and Radio Programs." *Racine Journal Times,* 7 Sept. 1955, p. 21.

Tribe, Lawrence. *No Requiem for the Space Age: The Apollo Moon Landings and American Popular Culture.* Oxford UP, 2014.

Ward, Bob. *Dr. Space: The Life of Wernher von Braun.* Naval Institute Press, 2005.

Wolters, Larry. "Mr. Disney Shows Way to the Moon." *Chicago Tribune,* 29 Dec. 1955, p. 14.

Wright, Mike. 1993. "The Disney-Von Braun Collaboration and Its Influence on Space Exploration." *Marshall History.* https://www.nasa.gov/centers/marshall/history/vonbraun/disney_article.html.

4

The Launch of Apollo 11

A Necessary Mission in the Space Race vs. Superfluous Government Spending

PATRICIA ROSSI

On July 16, 1969, in Cape Canaveral, Florida, the triumphant launch of Apollo 11 marked a historical moment indelibly etched in the memory of an entire generation. More than fifty years ago, with wild anticipation, extended families and neighbors gathered around their television sets to watch the launch. According to NASA records, an estimated six hundred million people tuned in and witnessed a watershed moment in the world's history. This monumental occurrence would indeed be documented and shared with future generations, captured in grammar school textbooks and high school social studies curricula as a proud and shining moment for the United States, with vibrant pictures of the historical blast-off, a detailed discussion of the United States' Space Race with the Soviet Union, the immortal words of Neil Armstrong, the American flag staked in lunar soil—certainly all pivotal moments in United States history—included in class lectures for the past half-century. Often accompanying these history lessons are pertinent snippets from President John F. Kennedy's speeches made in the early 1960s, a recording of his emphatic proclamation that by decade's end American astronauts would indeed land on the Moon.

The noteworthy success of Apollo 11, in conjunction with the other oft-cited accomplishments of the 1960s space program, has been applauded and characterized as a crowning achievement, taught and celebrated in classrooms across the nation for decades. However, rarely presented to the students are the historical details with respect to the American public's reaction, specifically the organized protests and dissension that challenged the 1960s Space Race, including the Apollo 11 launching and expedition.

The opposition was founded in a firm conviction that the government was recklessly spending exorbitant funds on the space program at the expense of individuals residing in underserved communities in need of food, quality housing, improved educational systems, and a means to find and sustain employment (Paul, Moss 218).

Lessons typically presented in the classroom have been remiss in fully exploring and discussing the extent of and rationale for the opposition to the government's space funding allocation, most especially the protests and the dissent generated by the African American community. The National Education Association website provides age-appropriate lessons regarding the history of Apollo 11 for students K-12, however when perusing the suggested curricula offered, none include a discussion concerning the protests voiced by the African American community. A detailed examination of the documented criticism of the 1960s space program, with specific focus on the launch and expedition of Apollo 11, provides a distinct, multi-faceted educational opportunity to illustrate an essential connection between government space funding and racial politics: to understand the protests as an integral component of the much broader civil rights movement; to highlight the leadership of Martin Luther King Jr. as well as Reverend Ralph Abernathy (civil rights activist and mentor to King); to credit the response of the Johnson Administration, via federal legislation, as instrumental in advancing the cause and concerns expressed by the African American community; to study the shift in the political landscape with respect to NASA funding; to accept the existence of two diverse contentions—fervent American support for government space funding versus flagrant disapproval via organized protests; and finally, to embrace the concept that these diametrically opposed contentions ultimately cultivated a mutually beneficial relationship instrumental in advancing their respective causes.

With respect to the launch of Apollo 11 and opposition from the African American community, a study of the events on the eve of Apollo 11's launch clearly illustrates the proposed historical link between government space funding and race relations. Imagine this remarkable scene, as it unfolded: It was July 15, 1969, and Reverend Ralph Abernathy arrived at Cape Canaveral accompanied by twenty-five poor southern African American families and four mules pulling two tottering wagons. Surrounded by newspaper reporters and television crews, Abernathy introduced himself and his entourage as members of the Southern Christian Leadership Conference Poor People's Campaign. Abernathy's group marched in peace, hands joined, singing the hymn "We Shall Overcome." In the distance, Apollo 11's Saturn V rocket was majestically positioned on its launch pad, quite a contrast to the wood wagon and four mules. Abernathy and his followers were painfully cognizant of NASA'S expenditures, most especially with respect to the Apollo 11 expedition, as newspaper headlines consistently reported that the government had spent an estimated $25 billion for said expedition. Abernathy addressed the crowd, which included NASA's top administrator, Thomas O. Paine, and stated that he and members of the Southern Christian Leadership Conference had traveled to Apollo 11's launch site not to specifically protest the government's space program and the imminent launch and expedition but rather to seek an explanation for "the arithmetic, the spending of billions for space and pennies for hunger."

Abernathy assembled his followers in front of the NASA delegation and proclaimed, "one-fifth of the population lacks sufficient supplies of food, clothing, medical care, and shelter. My people are Americans too, with no homeland but America, and we must improve their lot" (Weinraub 22). Initially, Abernathy acknowledged the accomplishments of NASA. He stated, "On the eve of man's noblest venture, I am profoundly moved by the nation's achievements in space and the heroism of the three men embarking on the Moon." Abernathy referred to the launch site as "holy ground," but quickly qualified his description and added, "it will be even more holy once we feed the hungry and care for the sick and provide for those who do not have homes." Abernathy emphasized his presence and his purpose, "to protest the country's distorted sense of national priorities" and questioned "the inexcusable gulf that exists between America's technological abilities and our social injustices." Abernathy continued, "It is a grave inequality to spend billions to put human beings on the Moon when the nation refuses to allocate sufficient funds to prevent starvation of human beings on Earth. America has reached out to the stars but not to the starving poor." Abernathy then entertained the feasibility of future space exploration and added, "we may go to Mars and Jupiter and to the heavens beyond, but as racism and poverty prevail we as a civilized nation have failed." Abernathy intently looked at Paine and concluded with a specific plea: "I want NASA scientists and engineers and technicians to find ways to use their skills to tackle the problems we face in society."

Paine then stated, "If we could solve the problems of poverty in the United States by not pushing the button to launch men to the Moon tomorrow, then we would not push that button." With apparent humility,

Paine admitted, "the technology inherent in Apollo spacecraft is child's play compared to the tremendously difficult human problems with which you and your people are concerned" (Maher 12). Paine offered a suggestion: "the space program and the benefits of science could perhaps be used to solve the problems of the poor. The space triumphs will make a resolution for Americans to band together to fight for the programs you are talking about." Paine offered a figurative and literal invitation to the protesters when he stated, "I hope you will hitch your mule wagons to our rockets and use the space program as a spur to the nation to tackle problems boldly in other areas." He then distributed to Abernathy and his group VIP passes to the launch. Paine concluded with a request and a promise as he asked Abernathy and his accompanying group "to mention the three astronauts in their prayers" and then "pledges to do what he can to adapt food concentrates developed in the space program to aid the Nation's malnourished" (Greider A1). In a final response to Paine's words, Abernathy held up a mock noose and gestured toward Paine's neck. Paine slowly bent forward as Abernathy carefully slipped the rope over Paine's head. Attached to the mock noose was a handwritten sign in bold lettering: "I helped hang poverty." Paine's response to Abernathy's concerns voiced on behalf of the African American community was perceived by the general public as quite sincere and sympathetic. Most certainly, Abernathy was successful in his endeavors to orchestrate a dramatic confrontation on the eve of the launch, and he captured worldwide attention for his cause and his stated concerns for the African American community (Maher 49).

Several hours before Abernathy and members of his coalition arrived at Cape Canaveral and confronted Thomas Paine, Abernathy led a prayer service at St. Paul's Church, in a predominantly African American neighborhood in West Cocoa, Florida. The church, narrow and red-bricked, was overflowing with hundreds of parishioners, as well as television and radio reporters. Hosea Williams, a co-organizer of the Poor People's Campaign and co-leader of the Southern Christian Leadership Conference, clearly delineated the intended goal with respect to marching to and protesting at Cape Canaveral. "We don't want to be obstructive here; we don't want to irritate. We are spending billions of dollars to explore outer space, but if Americans spent the same amount of money feeding the poor and hungry, then poverty and hunger would be gone from the face of America today." Williams added, "Our intention is to protest our nation's inability to choose humane priorities." Williams then introduced Lilly Belle Holt and her nine children from Social Circle, Georgia, to the gathered news reporters and shouted, "this woman gets eighty-two dollars a month and a one-room shack." Pointing at Holt and her children, Williams asked, "Why should we be worrying about sending three men to the Moon when here are ten people dying of starvation? If we can spend a hundred dollars a mile to send three men to the Moon, can't we, for God's sake, feed our hungry?" (Greider A1).

On July 20, 1969, the day of the Moon landing, the *New York Times* published an article written by Bernard Weinraub titled "At the Launching, Turbulent America in Microcosm." In the article Weinraub first describes the majestic launch pad, the VIP stands at the Kennedy Space Center overflowing with attending dignitaries, Hollywood actors, well-dressed vacationing families, and scholars, but he quickly informs the reader that only a few miles away are peeling wooden shacks inhabited predominantly by African Americans. Weinraub then introduces the reader to "a Mrs. Lucille Wade, the mother of six children who went hungry." Weinraub shares a quote from Wade: "I tell you I try. When the end of the month comes, my three in school just eat that one meal they give them and that's all. I give my others grits and margarine for breakfast. I don't fault anyone" (E1). In a *New York Times* article published July 20, 1994, commemorating the twenty-fifth anniversary of the Moon landing, the reporter, John Tierney, much like Weinraub, describes the dichotomy via reference to an archived *New York Times* editorial published in July 1969. Tierney writes, "The moonwalk was a polarizing event, greeted by street celebrations of Yankee ingenuity and by protest demonstrations against American wastefulness." The ambivalence was reflected in the *New York Times* coverage, which featured its largest headline ever, photographs of enraptured crowds watching television screens in Central Park and Trafalgar Square, and an editorial hailing "A new era leading to realms beyond comprehension." But Tierney further shares with the reader that his reference to the 1969 *New York Times* editorial "also warned as man gains another world—with the prospect of adding another and still another, literally ad infinitum—he stands in imminent danger of losing his own" (A1).

Inspired by Abernathy, emulating and following in his footsteps were dozens of civil rights activists who orchestrated boycotts, sit-ins, and demonstrations; consequently, a solidarity emerged across the nation decrying the misappropriation of exorbitant funds for space exploration at the expense of impoverished Americans and crumbling inner-city neighborhoods. On

July 20, 1969, at NASA's Mission Control Center in Houston, Texas, a group of approximately seventy-five African Americans gathered, identifying themselves as members of the Houston Welfare Rights Organization. The assembled group convened around a small-scale model of the Eagle lunar landing module and peacefully demanded that further lunar missions be ceased immediately until the federal government actively engaged in viable measures to provide affordable and healthy housing for those individuals living in inner-city neighborhoods ("Hunger Protests Held at NASA" 5).

As the fervor, support, and enthusiasm for Apollo 11's returning astronauts dramatically increased, so too did the unified dissension and the accompanying demonstrations with respect to superfluous government space funding. When President Richard Nixon declared July 21, 1969, as a federal holiday, proclaiming it as a day to celebrate Apollo 11's success, encouraging all Americans to participate in parades and government-sponsored festivities, Marion Barry—then co-founder of Pride Inc., a program with a stated mission to assist unemployed Black men to secure gainful employment—responded with an organized protest. Barry encouraged the country's African American community to reject Nixon's proclamation and refrain from partaking in the nation's scheduled celebrations. Barry recommended that the African American communities celebrate July 21, 1969, by creating inner-city coalitions and addressing urban renewal and environmental concerns. Barry posed the question, "Why should Blacks rejoice when two White Americans land on the Moon when White America's money and technology have not yet reached the inner city? Why should Blacks celebrate Monday as a national holiday when President Nixon did not feel that Dr. Martin Luther King's assassination deserved to be observed?" (Ezell 20; Paka 20).

On July 22, 1969, the *New York Times* published an article titled, "In New York, Fatigue for TV Watchers, but Some Work Was Done." The article describes the thousands of New Yorkers, mesmerized, up all night watching the Apollo 11 coverage on their television sets. The article also provides details with respect to the President's proclamation: "All but essential city and state offices were closed, as were the New York and American Stock Exchanges. Schools were shut down and many businesses suspended operations for the day." It is quite worth noting that the article featured a contingent of employees who performed their work duties in protest of the proclamation made by President Nixon. The article states, "Some workers who were offered the day off refused to stay at home. The offices of the

Coney Island Associates Corporation, which handles $600,000 of antipoverty programs in the area, stayed open as a protest against the huge spending on the space program. All but three or four members of the 250-person staff came to work, and most refrained from listening to the radio or watching the television coverage of the day's lunar activity." The article concludes with a comment from James McGrath, a spokesman for the agency: "That shot to the Moon could have paid to clean up the problems of Coney Island, and the next shot could help Appalachia" (1).

Akin to Nixon's declared federal holiday, in response to the plethora of ornate galas and formal events held in honor of the Apollo 11 astronauts, Aldrin, Armstrong, and Collins, civil rights leaders were further prompted to engage in numerous acts of civil disobedience. On August 13, 1969, an elaborate dinner was held at the Los Angeles Century Plaza Hotel, with a guest list of more than fifteen hundred, including governors, cabinet members, members of Congress, Supreme Court Justices, a wealth of foreign diplomats, and Hollywood actors. Additionally, President Nixon was scheduled to award the medal of freedom, the highest honor presented to a civilian by the United States, to Aldrin, Armstrong, and Collins. As wine was sipped, pleasantries exchanged, and the astronauts toasted and formally honored, demonstrations ensued in the Century Plaza Hotel parking lot. Leaflets were distributed with the proclamation, "Americans have moved forward in space, [but] we have repeatedly lagged behind in areas of civil rights and intolerable ghetto conditions" (Oberdorfer A1). The printed message was strewn about, affixed to car windshields, and handed to gala invitees as they departed and thus serves as a tangible illustration of the historical connection between government funding of space exploration and racial politics. A *New York Times* article published on August 15, 1969, titled "Astronauts Find Mixed Reactions: The Uninvited Hold Protest as Diners Hail Crew," describes in vivid detail the crowd that gathered outside the Century Plaza Hotel: "The outsiders wore jeans, not formal dresses, work shirts, not tuxedos, sneakers, not patent leather pumps. They too, akin to Armstrong, talked about a 'new era.' They chanted in unison. They wondered if it was right to spend money to reach the Moon while people went hungry. They wondered why scientists could build a rocket and not keep the air and the rivers clean" (Roberts 14).

There were also less structured but nonetheless effective boycotts. On the day Apollo 11 was launched, witnessed by millions, and celebrated as a laudatory event, African Americans gathered in bars and restaurants in predominantly African American neighborhoods throughout the country, and the patrons purposefully turned the television dials to local sporting events. At the Metropole Bar on 23rd street and Michigan Avenue in Chicago, as well as at pubs in Harlem, a local baseball game was shown in lieu of the lunar mission. At the Harlem Cultural Festival at Mount Morris Park in New York, attended by an estimated fifty thousand African Americans, when the music was momentarily stopped to announce the successful lunar touchdown, boisterous boos and shouts of discontent filled the air ("Talk of the Town" 26). The *New York Times* printed an article on July 27, 1969, titled, "Blacks and Apollo: Most Couldn't Have Cared Less," detailing African Americans' reaction to the Moon landing. In this article, a Miss Sylvia Drew, identified as an attorney for the NAACP Legal Defense and Education Fund, states: "This proves that white America will do whatever it is committed to doing; if America fails to end discrimination, hunger, and malnutrition then we must conclude that America is not committed to ending discrimination, hunger and malnutrition. Walking on the Moon proves that we do what we want to do as a nation." A Miss Victoria Mares, director of a poverty program in Saginaw, Michigan, is also quoted in the same *Times* article: "If we compare the United States to a man who has a large family, they have no shoes, no clothing, no food, and the rent is overdue. But when he gets paid, he runs out and buys himself a set—another set—of electric trains. We are supporting our government's hobby at the expense of its poor citizens." Roy Wilkins, the NAACP Executive Director, is also quoted as saying, "The Moon shot is a cause for shame; there's something wrong with the Government's priority system" (Johnson E6).

Earlier in the decade, in a much more formal setting, a viable link between government space funding and race relations was also quite apparent. Dr. Martin Luther King Jr., accompanied by fellow SCLC leaders Andrew Young and Walter Fauntroy, testified before a Senate subcommittee investigating the plight of inner cities. King stated, "There is a striking absurdity in committing billions to reach the Moon, where no people live and from which none can presently benefit, while the densely populated slums are allocated minuscule appropriations." King succinctly conveyed his concern that the space program was diverting funds from creating and implementing government programs to improve living conditions in inner city neighborhoods: "With the continuation of these strange values in a few years we can be assured that we will set a man on the Moon and with an adequate telescope he will be able to see the slums on earth with their intensified congestion, decay, and turbulence" (Maher 22).

A surge of celebratory memorabilia—space toys for the young, aspiring astronaut; posters; glossy magazines boasting interviews with Aldrin, Armstrong, and Collins—all flooded mainstream media. Simultaneously, voices of discontent and criticisms were expressed via African American newspapers and magazines. A plethora of interviews, political cartoons, and editorials expressed opposition to space exploration, contending that funds expended for space are to the detriment of impoverished inner-city inhabitants. An overwhelming majority of these editorials acknowledged the government's dedication and determination in the achievement of the Apollo 11 Moon landing but expressed extreme disappointment and discontent with respect to the extravagant funds allocated, citing the existence of deplorable living conditions and substandard educational systems here on Earth, mostly in neighborhoods populated by African Americans. On July 26, 1969, a poll published in the *Michigan Chronicle* reported that African Americans concur that although "the feat was awesome, the financial expenditures were too enormous, [and] the funds should have been put to use in curing earthly ills" ("Billions for Trip" A8). In the July 24, 1969, issue of Oklahoma City's *Black Dispatch*, an editorial describes the space landing as a "great human achievement" and recognizes that "a great step forward has been made." However, the writer qualifies said achievement, sharing with the reader that "more than 15 percent of our citizens are living at or below poverty level" ("Comments Vary" 16).

In its July 26, 1969, edition, the *Michigan Chronicle* printed an editorial that congratulates NASA for its "ingenuity, determination and open purses of the American people." The writer then reminds the reader that Apollo 11 was achieved against unfavorable odds, pointing out that "just a few years back the United States found itself lagging far behind the Soviet Union." The editorial labels this success as sheer irony, writing "we could surpass the Russians and win the Moon race, while we find ourselves unable to wipe out poverty, disease, and hunger right here on this hemisphere" ("Billions for Trip" A8). The *Iowa Bystander*, on July 24, 1969, shared the hopeful sentiments of a writer who states "we are at the beginning of a most exciting, beautiful era, but we should now face up to the task of providing in spirit, mind, and body for those less fortunate earthlings" ("Letters from Our Readers" 6). In yet another editorial printed in the *Michigan*

Chronicle, Pastor James E. Wadsworth Jr. applauds the "tremendous technological achievement" but ponders why "a nation able to plan and have the know-how to get men on the Moon was unable to plan with the same zeal and national commitment to solve our problems of poverty, poor public education, crime, slum dwellings, etc. . . ." Pastor Wadsworth provides the reader with an answer: "The truth is this country could, but we do not have the commitment from the national political and business leadership." Pastor Wadsworth cautions the reader against "getting caught up in this Moon landing hysteria, because we have to live on this globe and there is much that needs to be done to make our existence here worthwhile" ("Brothers Assess" A9).

On August 2, 1969, the *New Pittsburgh Courier* featured an editorial that echoes the same sentiments expressed in dozens of other African American papers nationwide; the writer recognizes "the conquest of the Moon as the affirmation of man's infinite capacity to overcome all obstacles in his path to knowledge and technological accomplishments. The success of the astronauts establishes beyond the peradventure of doubt man's superiority over the blind forces of nature." The writer concludes with the following question: "If these extremely difficult and complex tasks can be performed, why can't cities be cleaned up and made to work . . . why can't the races live together and make brotherhood a joyful reality?" ("Moon Conquest" 8). Two months after the Apollo 11 mission, *Ebony* magazine published a rather reflective and insightful editorial. Akin to the editorials cited above and commencing with initial praise for the accomplishments of the astronauts and the mission, the writer suggests, "Man's space frontiers are limitless, and perhaps it is only a matter of time before man journeys to Mars and Venus and beyond." The writer explores an interesting aspect of potential future space missions and asks, "As space scientists continue to explore the universe, what do they plan to say to any intelligent being they might find on Mars or Venus or any of the millions of the stars and planets in this universe?" The writer answers his own question: "We are from planet Earth, we have millions of people starving to death back home, we spend more on space exploration than on housing, education, culture and perhaps food" ("Giant Leap" 58).

A Dr. Benjamin W. Watkins, crowned as the honorary "Mayor of Harlem," was quoted in the *Negro Weekly* and the *Amsterdam News*: "Money could have best been spent on the rehabilitation of cities. We in Harlem are demanding that the trio of astronauts include in their itinerary Harlem,

Watts, or some other ghetto, rather than Moscow or England." The *Amsterdam News* carried an editorial whose last, rather poignant sentence reads: "Yesterday the Moon, tomorrow maybe us" (Johnson E6). These concerns of African Americans were also expressed through poetry. Gil Scott-Heron's poem, "Whitey on The Moon," gained instant popularity within the African American community and was subsequently set to music, released by a record company, and broadcast on the airwaves. The words truly capture the African American perspective on the Moon landing: "A rat done bit my sister Nell with Whitey on the Moon . . . Her face and arms began to swell . . . Whitey on the Moon . . . I can't pay no doctor bill, but Whitey on the Moon. Ten year from now I'll be payin' still . . . Whitey on the Moon." The words convey the prevailing sentiments within the African American community: blatant disregard for the African American's health, well-being, and living conditions in favor of space exploration (Maher 30).

The existence of a historical link between government space funding and race relations is also evidenced by a measurable shift in the political landscape. Initially, a majority of government representatives were firm proponents of lunar funding while a minority expressed complete disdain and opposition. However, a new contingency of outspoken politicians began to emerge, characterized by an intentional withdrawal of initial support for lunar funding, a requalification of the previously stated positions, and ultimately a wholehearted embrace of the concerns voiced by the African American community. In 1969, Charles Grigsby, a newly elected representative serving on an urban renewal committee in Boston, adamantly stated at a well-attended city hearing that "government funds should have been allocated to fight Boston's urban housing crisis, not land a man on the Moon" (Chen 54). Former New York City Mayor Ed Koch, who served as a Congressman from New York during the space era, stated, "I cannot justify approving moneys to find out whether or not there is some microbe on Mars, when in fact I know there are rats in Harlem apartments" (Koch 82). Senate Majority Leader Michael Mansfield, a Democrat from Montana, concurred: "While I honor the accomplishments of United States astronauts, we have a lot of problems here on Earth that we have to face up to. When we solve those problems we can consider further space efforts. The needs of the people on Earth and especially this country should be a priority" (Carter A1).

Senator Edward Kennedy, formerly a notoriously staunch supporter of NASA and proclaimed advocate of space exploration funding, candidly withdrew his support. On May 19, 1969, he was the featured speaker at a library dedication ceremony at Clark University. The honoree was Robert H. Goddard, a NASA physicist who launched the nation's first liquid-fueled rocket. A crowd of more than three thousand were in attendance. Kennedy's speech clearly qualified his support of NASA: "I believe in the space program, but we also have other national goals including the abolition of poverty and a decent house for every American." Kennedy recommended "the diversion of a substantial portion of the space budget to solve more pressing problems on Earth" and shared his vision: "Our challenge today is to use the same techniques and the same discipline of the Apollo mission to lower the costs of production for home-building." On May 20, 1969, the *New York Times* published an article by Robert Reinhold titled, "Kennedy Puts Earth Needs Ahead of Space Program." Reinhold reported on the library dedication and reaction to Kennedy's speech. Reinhold wrote, "It was an unlikely platform the Senator had chosen to revive the space controversy as he spoke at the dedication of the futuristic $5.4 million Robert Hutchings Goddard Library at Clark University." Reinhold quotes Kennedy: "We need not try to get to Mars or Venus merely because the Russians might get there. I am for the space program, but I want to see it in its right priority." Reinhold shares with the reader, "not everyone at the ceremony was pleased by the Senator's words; the guests included Colonel Edwin Aldrin Jr. of the Air Force, an astronaut who is likely to fulfill Professor Goddard's long-harbored dream of reaching the Moon by becoming part of the crew of July's mission, and Dr. Wernher von Braun, the rocket expert." Reinhold includes von Braun's admonition, "I hope there will not be too much budget cutting; if you want to stay in the space business there is a certain minimum level." Reinhold then shares with the reader that when questioned by reporters, Colonel Aldrin refused to comment, as did NASA officials when questioned about Kennedy's speech. The article concludes with a statement released by Kennedy's Senate Office following the dedication ceremony. Most certainly in an attempt to quell the negative reaction to Kennedy's comments, his office wrote, "Mr. Kennedy was not calling for the slowdown after this July's planned lunar landing by Apollo 11. In fact, he wants the flight planned to explore the lunar surface to go ahead as scheduled" (Reinhold 1).

The political shift in space funding support was also evidenced on the Senate floor. The eruption of heated debates between senators with respect to NASA's multi-billion-dollar proposed budget was a testament to the changing political tide. Some Senators argued that the budget should be slashed, with monies allocated to city programs and to addressing urban needs. Wisconsin Democrat William Proxmire asked his colleagues on the Senate floor, "Why do we have to keep going to the Moon? Man has immediate needs on Earth that need taking care of." One noted debate lasted seven hours, resulting in funds not being reduced for NASA. By a vote of 38 to 35, the Senate rejected a measure that would have cut $122 million from the $3.3 billion NASA appropriation approved previously in committee. The same day that the Senate refused to cut space spending, it also refused to increase it. An amendment to boost NASA funds offered by Senator Goldwater was defeated 58 to 15. Certainly, the intensity and longevity of the infamous seven-hour debate—and the tally of votes both in favor and against reduction of NASA funding and allocated increased spending to NASA—serve to illustrate a political trend marking the beginning of steep reductions in federal funding for the space program (Carter A1).

The words spoken and actions taken by President Lyndon B. Johnson during and after his administration are also illustrative of the historical link between government space race funding and racial politics. Johnson's response and political maneuvers, and the legislation implemented during his term, are pivotal in elucidating a connection, ultimately a rather unexpected symmetry between space funding and race relations. In 1958, Johnson served as chairman of the Special Committee on Space and Aeronautics, where he supervised the creation of NASA. Johnson later worked to achieve Kennedy's goal of landing a man on the Moon by decade's end. Simultaneously, Johnson extended his efforts to fight poverty and discrimination, as he was also cognizant of inner-city issues. On January 8, 1964, Johnson declared "an unconditional war on poverty in America." At the height of the space race, Johnson implemented federal legislation that created anti-poverty programs, including Medicare and Medicaid, Head Start, and Upward Bound. Johnson was a firm advocate for urban renewal and established community programs to assist minorities in achieving economic self-sufficiency. When civil rights activists criticized the space program for not hiring African American staff, including astronauts, Johnson responded to this and other similar examples of discrimination by implementing federal legislation, specifically the Civil Rights Act of 1964, which prohibited discrimination based on race. However, the same legislation promulgated by Johnson included a clause exempting federal agency compliance. In 1965, Johnson responded to the criticism expressed with respect to the federal exemption by enacting a law which also prohibited racial discrimination in the workplace, including employees of the federal government. Johnson adeptly balanced and gingerly tread between the proposed needs of NASA and his acute awareness and concern for impoverished Americans. Johnson's intentional navigation between space funding allocation and race relations demonstrates an inherent connection between the two. This is especially evident in Johnson's declaration that the space program was in fact diverting funds from his enacted welfare program. Johnson advocated reductions in NASA's budget and took further action in his support of poverty-stricken Americans when he affirmatively stated that reallocating funds from the space program was not enough. Johnson instructed NASA that they were to use their space technology to combat urban environmental issues, urban pollution, and waste. NASA engaged in studies and produced a plethora of information and suggestions responding to Johnson (Maher 36–37).

Johnson, post-administration, delivered a speech at Cape Canaveral the day before the Apollo 11 launch. He was invited to speak at a luncheon in honor of NASA's second chief administrator, James Webb. Johnson's carefully chosen words to the invitees further demonstrate the historical link between 1960s space funding allocation and racial politics: "If we can lead the world to the Moon, we can lead them to peace and bountiful prosperity at home." Johnson further stated that while "the space race had made a "valiant and very proud record for our country, the nation has much to do in education, poverty and other areas affecting American cities" ("Hails NASA" 5). Johnson took care to honor and recognize the achievements of the space program and the impending launch of Apollo 11, but he also expressed concerns for the homeless and those living in deplorable inner-city neighborhoods. It is worth noting that as Johnson skillfully wove and proclaimed his concerns for both the space program and the impoverished American, literally down the road, at the same hour, stood Reverend Ralph Abernathy, voicing the very same concerns. Two historical figures from very different worlds were synchronized in their message intended for the masses. While both acknowledged the accomplishments of NASA, they also questioned the appropriation of space funds in a country where many people are starving, homeless, and subjected to a substandard educational system (Abernathy 294–295). Clearly, the echoed sentiments of Johnson and Abernathy validate the historical link between government space funding appropriation and race relations.

Moreover, the well-established and respected Gallup poll provides valuable statistical data with respect to the American public's reaction to funding the space program in the early 1960s, as well as the possibility of man going beyond the Moon to Mars. The article written by George Gallup, printed in the *Washington Post* on August 7, 1969, and titled "Public Cool to Manned Mars Landing: The Gallup Poll," first explores the public's reaction to the question, "How would you feel about landing a man on the planet Mars—would you favor or oppose the United States setting aside money for such a project?" The reported results indicate that the overwhelming majority opposed such funding. One of the poll participants, a college instructor from Lansing, Michigan, is quoted: "With all the poverty, crime, [and] urban decay that we have on this planet, I see no reason why we should use all of our resources to get to a planet where life probably does not exist." With respect to the minority, Gallup writes, "those in favor reason that the scientific exploration of space must be continued to advance the knowledge of mankind and to stay ahead of Russia." The article concludes on a rather significant and revealing note. Gallup states, "The public was also lukewarm about proposals in the early 1960s to begin an extensive program designed to land a man on the Moon." Gallup shares that in May 1961, when President Kennedy requested that Congress increase expenditures for the space program in an effort to land a man on the Moon before Russia, "only one person in three were willing to see the United States spend the billions necessary to get a man on the Moon." Gallup's shared findings indicate and confirm that the position held by the African American community with respect to space funding was certainly more widely accepted than perhaps ever indicated (F4).

In conclusion, this research establishes the existence of a historical link between government funding for the 1960s space program and race relations, leading to a richer understanding of the politics inherent in the space race. The lesson that emerges is multi-faceted: deeply woven within the tapestry of the Apollo 11 expedition and the 1960s space program are threads tethered to racial politics and the history of the civil rights movement. Moreover, it is ironic that the space race may have been instrumental in advancing the causes of the African American civil rights movement. How? The reaction to government expenditures on space exploration provided a platform for dissension, the opportunity to protest, and thereby to cast a spotlight on the racial inequalities present—the deplorable living conditions endured by Americans—most especially African Americans.

The enormous government expenditures required for the technological achievements of the space race and the success in accomplishing space goals provided a sharp and undeniable contrast to the limited government resources and lack of progress in solving the problems of poverty, pollution, and racial inequality. Having allocated so many resources to successfully putting men on the Moon, the federal government had to respond to the voices protesting the neglect of societal problems in cities across America. Perhaps the broader, contemporary lesson is that we can better appreciate the accomplishments of the 1960s space program, including the Apollo 11 expedition, when accompanied by an understanding that protests of the space program were equally critical to advancing the American project. In reaching for the skies above and the crowning achievement of landing men on the Moon, American democracy, through peaceful dissension, also made a giant leap for racial equality back home on Earth.

Works Cited

Abernathy, Ralph David. *And the Walls Came Tumbling Down. An Autobiography*, Lawrence Hills Books, 1989.

"Billions for Trip to the Moon: Now How About Care for Poor." *Michigan Chronicle*, 26 July 1969.

"Brothers Assess Moon Shot." *Michigan Chronicle*, 9 August 1969.

Carter, Philip. "NASA Cuts Rejected by Senate: Senate Beats Attempts to Trim Space Funds." *Washington Post*, 8 July 1970.

Chen, Victor. "Some Attack Cost of Trip." *Boston Sunday Globe*, 13 July 1969.

"Comments Vary on Outer Space Feat." *Black Dispatch* [Oklahoma City, OK], 24 July 1969.

Ezell, Edward. "Apollo: So what? Earth Turmoil Dims Triumph." *Daily News* [Williamson, WV], 19 July 1979.

Gallup, George. "Public Cool to Manned Mars Land: The Gallup Poll." *Washington Post*, 7 August 1969.

"Giant Leap for Mankind" *Ebony*, Sept. 1969.

Greider, William. "Protesters, VIPS, Flood Cape Area." *Washington Post*, 7 July 1969.

"Hails NASA: LBJ Arrives at Cape." *Chicago Sun-Times*, 16 July 1969.

"Hunger Protest Held at NASA: Welfare Group Sits by LM Mock." *Toledo Blade* [Toledo, OH], 21 July 1969.

"In New York Fatigue for TV Watchers, But Some Work Was Done." *New York Times*, 22 July 1969.

Johnson, Thomas A. "Blacks and Apollo, Most Couldn't Have Cared Less." *New York Times*, 7 July 1969.

Koch, Edward. "Science: Post Mortem on Apollo." *Time*, 4 May 1970.

"Letters from Our Readers." *Iowa Bystander*, 24 July 1969.

Maher, Neil. *Apollo In the Age of Aquarius*. Harvard UP, 2017.

"Moon Conquest." *New Pittsburgh Courier*. Pittsburgh, PA., 2 August 1969.

Oberdorfer, Don. "Apollo Astronauts Hailed by Millions." *Washington Post*, 14 August 1969.

Paul, Richard, and Moss, Steven. *We Could Not Fail: The First African Americans in the Space Program*. University of Texas Press, 2015.

Paka, Vincent. "Barry Slams Apollo 11 Mission." *Washington Post*, 7 July 1969.

Reinhold, Robert. "Kennedy Puts Earth Needs Ahead of Space Program." *New York Times*, 20 May 1969.

Roberts, Steven V. "Astronauts Find Mixed Reactions: The Uninvited Hold Protests as Diners Hail Crew." *New York Times*, 15 August 1969.

"The Talk of the Town: The Moon Hours." *New Yorker*, 26 July 1969.

Tierney, John. "Earthly Worries Supplant Euphoria of Moon Shots." *New York Times*, 7 July 1994.

Weinraub, Bernard. "At the Launching Turbulent America in Microcosm." *New York Times*, 20 July 1969.

Weinraub, Bernard. "Hundreds of Thousands Flock to be There." *New York Times*, 16 July 1969.

5

Scientists Without Borders

Immigrants in NASA and the Apollo Program

ROSANNA PEROTTI

That Americans were first to walk on the Moon was particularly apt.[1] As the world's top immigrant destination for the past fifty years, the United States best fits the descriptor of a nation of immigrants, travelers who journeyed beyond the gravitational pull of their home country, facing risk and adversity to forge new paths of discovery. Space travel itself shares many of the characteristics of the act of immigration. Immigrants and the children of immigrants played an outsized role in the development of Apollo and many subsequent American space endeavors. Indeed, the entire field of space travel continues to be indebted to the contributions of all sorts of immigrants to the United States: those who started as foreign students, as refugees, as employment-based immigrants, as the spouses and minor children of immigrants, as undocumented immigrants—people of all kinds who dreamed of space travel and who followed their talents and dreams into work with NASA or its contractors.

This chapter explores the importance of foreign-born workers to NASA and the United States' science and engineering workforce involved in spaceflight. It examines the theme of space travel as the embodiment of the immigrant ideal in the American imagination. Finally, it outlines in anecdotal fashion some of the continuing contributions of immigrants and their children to NASA's work. I take two lessons away from this brief examination of immigrants and space travel. First, immigrants have a special relationship with the space program for a number of reasons, both symbolic and practical. Second, as Americans continue to debate major changes to our nation's immigration system, we would be wise to take into account how those changes will affect the quality of our nation's space program, which has benefitted greatly from the efforts of immigrant engineers, scientists, and workers from every corner of the world.

Foreign-Born Workers in the Science and Engineering Workforce

Public statistics on the prevalence of foreign-born workers in NASA are difficult to find. Though NASA carefully tracks and publicly reports its hiring statistics for women and ethnic and racial minorities, it is not clear what percentage of the workforce is foreign-born. However, it is widely known that foreign-born scientists and engineers are a growing and critical part of the broader US science and engineering workforce. Among those working in science and engineering (S&E) occupations in the United States in 2015, about 41 percent of master's degree holders and 36 percent of doctorate holders were foreign-born. Their numbers are greatest in the fields of engineering, computer sciences, and math, where more than one-half of doctorate holders are foreign-born, reflective of a long-term trend. Both the number and the proportion of foreign-born workers in S&E occupations have risen since the 1990s (National Science Board 2018, 2020, 2022).

Today these foreign-born workers tend to have higher levels of education than their US native-born counterparts. Seventeen percent of foreign-born S&E workers have a doctorate, compared to 10 percent of US native-born individuals in these occupations. In most S&E occupations, the higher the degree level, the greater the proportion of the workforce who are foreign-born. At the time of doctorate receipt, 76 percent of foreign-born noncitizen recipients of US S&E doctorates planned to stay in the United States in 2015, this proportion rising since the 1980s (National Science Board 2018). Foreign-born individuals make up a disproportionate number of the truly outstanding intellectuals and researchers in the United States. Forty-two percent of all Nobel Prizes awarded between 1901 and 2015 went to individuals working in the United States; of those American Nobel laureates, nearly one-third were born outside this country (Witte). The foreign-born are also overrepresented among the highest-impact researchers in some of science's most cutting-edge fields such as nanoscience and nanotechnology (Walsh). The foreign-born are represented disproportionately among tech entrepreneurs: an estimated one-fourth of the founders of technology and engineering firms established in the period 1995–2005 were non-natives (National Academy of Sciences).

If they are evident broadly in the sciences, these trends are also evident in the space industry, which has faced the dual challenges of an aging workforce and a shortage of US natives interested in STEM careers. While citizenship is required to work for NASA as a civilian employee, naturalized citizens are welcome. Naturalized citizens may become permanent residents as refugees, through family reunification, or through work visas, though the citizenship process takes time. As is the case elsewhere in the scientific and engineering fields, their language skills, cultural competencies, and networks of social and professional ties help the foreign-born to mobilize information and technical expertise, facilitating transnational relationships and exchanges (Walsh 108). Osaka-born and MIT-educated Masahiro Ono, writing an occasional blog about his work as a research technologist at the Jet Propulsion Laboratory in Pasadena, California, reflected:

> Listing the country of birth of my colleagues is almost like listing all the countries on the planet—Austria, Australia, Brazil, Canada, China, Columbia, Egypt, France, Guatemala, India, Israel, Korea, Mexico, Norway, the Philippines, Russia, Spain, Taiwan, UK, Zimbabwe, and so on. You can hear any accent. You can find any religion. You can find any color of skin, hair, and eyes. Jet Propulsion Laboratory (JPL) is the embodiment of MLK's dream where men and women are not "judged by the color of their skin but by the content of their character."

Much of NASA's work is done by contractors who may not have citizenship requirements. As we will see in the body of this chapter, there is ample anecdotal evidence of the impact of foreign-born scientists, engineers, and even astronauts in NASA.

Space Travel and the Immigrant Experience

From NASA's very beginnings, immigrant engineers, scientists, and technicians lent their talent, labor, and technical skills to the space program. But space travel itself always represented more than a scientific endeavor. Human spaceflight was one of the "great dreams" of the 1960s, as space historian Valerie Neal reminds us, and as a "big idea," spaceflight relied heavily on American cultural narratives. The Apollo program (1963–1972) conjured the image of pioneering the frontier in the 1960s—exploration and discovery were indispensable to America's history and continuing redefinition, and Americans welcomed the frontier as a metaphor for space exploration (Neal 15). The shuttle program (1972–2011) echoed the narrative of Americans "going to work." As the Apollo missions were replaced by the space shuttle, NASA supporters and commentators depicted the shuttle crews with imagery associated with blue-collar labor: "astronaut repairmen made service calls in a vehicle often called a space truck" (37).

Both of these narratives—"pioneering the frontier" and "getting the job done"—are closely associated with a third narrative that was becoming deeply ingrained in American national identity in the 1960s: the myth of the United States as a nation of immigrants and of the immigrant as the backbone of America's egalitarian democracy. This American immigrant myth was not born in the nineteenth or even in the early twentieth century, when immigration was peaking and Congress struggled to impose limitations and quotas. The myth reached wide acceptance only in the early 1960s. It is no coincidence that John F. Kennedy presented the immigrant myth most succinctly in his pamphlet, *A Nation of Immigrants*, in 1963, as Kennedy was preparing to ask Congress to overhaul the nation's immigration laws. At the same time, his administration was pressing furiously to put a man on the Moon by the end of the decade, a central goal of the New Frontier. Interestingly, Kennedy's space proposals were a far more important policy priority for the administration than immigration reform (the latter was not accomplished until 1965, as we shall see later). But his articulation of the "nation of immigrants" narrative provided powerful imagery in support of the space program he championed from the start of his administration.

Kennedy's articulation of the complex immigration myth featured not just a welcoming America, but an idealized immigrant, united with others by little other than a common love of freedom. Ours was "a nation of people with the fresh memory of old traditions who dared to explore new frontiers, people eager to build lives for themselves in a spacious society that did not restrict their freedom of choice and action" (Kennedy 2). Citing Tocqueville, Kennedy noted that immigrants' very poverty made them more inclined toward egalitarian democracy. No arena of American life was untouched by the influence of immigrants, and immigrants themselves were paragons of self-reliance, ingenuity, entrepreneurship, and pioneer spirit. "It was the future and not the past to which he was compelled to address himself," Kennedy wrote, describing the motivations of the nineteenth-century immigrant.

> Except for the Negro slave, he could go anywhere and do anything his talents permitted. A sprawling continent lay before him, and he had only to weld it together by canals, by railroads and by roads . . . This has been the foundation of American inventiveness and ingenuity, of the multiplicity of new enterprises, and of the success in achieving the highest standard of living anywhere in the world. (8)

The space program was the next frontier in the natural progression toward excellence. It evoked not only the immigrant's capacity for adventure and discovery but also his practicality and capacity to work hard and tame his surroundings. From the time of the English settlers, who "fought a rugged land" in the words of Kennedy, immigrants had to overcome adversity to earn their fortunes and shape their environment. They had worked as artisans, provided cheap labor for American farms, factories, mills, and mines, and climbed the economic ladder to provide succeeding generations with educational opportunities. They had moved forward to get the job done. Launched under the motto "Going to Work in Space," the space shuttle was a vehicle that could deliver satellites and repair them in orbit, carry commercial payloads, and support a research laboratory (Neal 40). Astronauts would carry out their work all but rolling up their sleeves as builders and repair technicians, wielding robotic arms and power hand tools. Businesses could use the shuttle as a workhorse to launch satellites or develop manufacturing capabilities. All of this economic productivity in space could be expected to resonate with a nation whose increasingly diverse immigrant workforce was transitioning to a new economy. American society was reflected not only symbolically but practically in NASA's missions. They produced results that appeared almost impossibly ambitious. NASA represented excellence: the best work in the world.

Space travel also mirrored some of the risks and hardships of the immigrant experience. As the American public began questioning the nation's investment in space travel through the 1980s, advocates harked back to this part of the immigrant narrative. In the aftermath of the 1986 Challenger tragedy, the Report of the Advisory Committee on the Future of the US Space Program (1990) reminded Americans that acceptance and resilience in the face of failure were a part of America's pioneer and immigrant legacies:

> In a very real sense, the space program is analogous to the exploration and settlement of the new world. In this view, risk and sacrifice

are seen to be constant features of the American experience. There is a national heritage of risk-taking handed down from early explorers, immigrants, settlers, and adventurers. It is this element of our national character that is the wellspring of the U.S. space program.

Early NASA Engineers Who Contributed to the Apollo Project

From the beginning, space travel in the United States was bound up in the immigrant experience, an experience almost as varied as immigrants themselves, from refugees to adventurers to former Nazis. The rocket engineers who laid the groundwork for the Apollo program were brought mostly from Germany, where in the 1910s and 1920s, aeronautics was far more advanced than in the United States. Wernher von Braun, the most prominent of the German rocketeers, had joined the Society for Space Travel in Germany during the late 1920s and early 1930s and had been invited by the German military to write his doctoral dissertation while developing rockets. Von Braun built Germany's first rocket development program at Kummersdorf in 1934 and continued at Peenemünde and finally at the infamous Mittelwerk (Central Works) near Nordhausen, where Germany's rockets were mass-produced by concentration camp slave labor in advance of being used, often unsuccessfully but sometimes with deadly results, against the British and allies. When von Braun, his staff, and associates saved his rocket program from the Russians and turned themselves in to the Americans in 1945, many were made "special employees" of the Army at Fort Bliss, Texas. Later, they were transferred to the Army's Redstone Arsenal at Huntsville, Alabama, where they designed the Redstone and Jupiter ballistic missiles. When the newly formed National Aeronautics and Space Agency took over the arsenal in 1958, von Braun became the first director of NASA's Marshall Space Flight Center. Together with a number of his German-born staff, he had become a citizen. Von Braun helped develop the ballistic missiles that launched the first American satellite into orbit, propelled the first Mercury astronaut into space, and, under NASA, the Saturn V Booster that sent twelve astronauts to the Moon. In 1970, he went to Washington to become NASA's deputy associate administrator for planning and died in 1977 (Brinkley, Lundquist, Jacobsen, Neufeld, Piszkiewicz).

Among von Braun's associates was German-born Arthur Rudolph, the man who had overseen the slave labor at the Mittelwerk rocket factory. Ultimately, Rudolph became manager of the Saturn V project office until his retirement in 1968. Hundreds of transplanted rocket scientists, mostly German-born, formed the core of the Apollo team at Marshall Space Center. Most were denazified, though some were returned to Germany (Brinkley 80). Patrick Scheuermann, former NASA director, wrote,

> They were not only rocketeers. They had families and were eager to contribute to the cultural life of their new nation and community. Many were active in civic organizations; several were excellent musicians and founding members of the Huntsville Symphony Orchestra. (Lundquist vii)

The Germans, according to Charles Lundquist, "mentored many more Army employees who were new to rocketry . . . the Peenemünde alumni were rather thoroughly Americanized by their interactions with native citizens. The transplantation process was essentially complete" (4, 5).

How did Huntsville receive these German-born immigrants? According to available accounts, with great excitement. The Chamber of Commerce organized a barbecue party, advertised in the local paper, to welcome them as "Special Civic Guests" in 1950. It was a major event for the small town; local civic clubs were expected to cancel their meetings to free up time for local notables to attend the party at Big Spring Park. Huntsville welcomed the prospect of the economic boost that the new engineers would bring. But the arrival of the engineers exposed flaws in Huntsville's delicate social fabric. Huntsville was thoroughly racially segregated, and Blacks figured little in either the welcoming festivities or the economic benefits that would flow from the Marshall Center. The new immigrants did not appear to be openly opposed to the system of racial segregation, and the depth of the German rocketeers' involvement in the Nazi system of slavery and genocide remained hidden until much later (Laney, Neufeld, 2008).

The Peenemünde rocket scientists were not the first immigrants to lend their skills to the fledgling space program. As early as 1926, the eminent physicist Robert Millikan had sought to recruit talent from Europe to start an aerodynamics laboratory as part of an effort to raise the profile of the California Institute of Technology (Caltech) and create a thriving aviation industry in Southern California. By 1929, Millikan had succeeded in attracting the Hungarian aerodynamicist Theodore von Kármán to leave Germany, where he had sought to start his academic career, and ultimately direct the Guggenheim Aeronautical Laboratory at Caltech (GALCIT). Other scientists, many of them Jews like von Kármán, were leaving Europe to avoid becoming part of the fascist military revival, as Germany, Italy, the Soviet Union, and Japan sought to enlist the brightest scientists in the development of their air power. Caltech fought to snap them up. Von Kármán's graduate students, who also included a fair number of foreign-born researchers, began to conduct rocket experiments in Arroyo Seco, a dry canyon wash north of the Rose Bowl in Pasadena, away from Caltech's main campus. The JPL was born, with von Kármán as its first director.

Ironically, von Kármán's expertise, linguistic ability, and familiarity with German scientific research prompted the US Army to send him to Germany immediately after the war to "find out firsthand how far the Germans actually have gotten in research and development." As a result, von Kármán, a Hungarian Jew, was part of the first group of US scientists to view, collect, and translate the records of the secret institute in northern Germany that supplied research for the V-2 rockets, as von Braun and his colleagues themselves were being detained and interrogated by the American military in Garmisch (von Kármán 282).

And so, even before the Peenemünde group established roots in Huntsville, von Kármán's group had been transforming American studies of theoretical aerodynamics in Pasadena. By all accounts, von Kármán was a charismatic public lecturer, teacher, mentor, prolific researcher, and a builder of international cooperation and institutions. Precisely because of his European training, he strove to bridge the gap between pure scientists and engineers, arguing that American-trained engineers should have a strong background in physics, chemistry, and, for some, the social sciences. As Millikan had hoped, von Kármán showed particular openness to innovation—as evident in his encouragement of his graduate students in their rocket experiments. He spoke some seven languages and lectured widely throughout his career, recruiting talented graduate students as he went, a practice he had begun during his years as a young professor in Germany. "I felt from the first that it was a good idea to create an international atmosphere at Caltech," he wrote. " . . . I had introduced graduate engineering to Caltech and now I wanted a representative number of foreign students to carry it on. They add the salt of new ideas to a school and help carry advances in science around the world" (von Kármán 155). Von Kármán was sometimes criticized for his active efforts to recruit foreign students, but he persisted, arguing that ethnic diversity would ultimately strengthen his program.

> I recall that a very proper professor from Cambridge, Massachusetts, once visited my laboratory. After viewing the turbans, Oriental eyes, and dark skins, and listening to the strange dialects, he asked me point-blank why I was emphasizing foreign students. I was amused. "And perhaps I should hire only Navajo Indians?" I said. He kept very quiet after that. (156)

Von Kármán's large international network of friends and former students had spurred him initially to journey to the United States, where his lab became "a mecca for theoretical studies" in aerodynamics, a field he largely introduced to the United States (Hanle xi). His passion for international collaboration kept him organizing conferences and organizations such as the International Academy of Astronautics, and later the postwar Air Force Scientific Advisory Group and NATO's Advisory Group on Aeronautical Research and Development (Thomas 1960). He was convinced that broadening the scientific conversation would lead not only to good science but also to political cooperation. "In my opinion, the only cure for world tensions is international thinking," von Kármán wrote (350).

The JPL developed rocket technology and missile systems in the 1940s. It built Explorer 1, the country's first satellite, as the United States was reeling from the Soviet Union's Sputnik launch in 1957. The lab was transferred to NASA upon the space agency's establishment in 1958. The JPL went on to send spacecraft to every planet in the solar system. Von Kármán was succeeded in the directorship by his star graduate student Frank J. Malina, and by two engineers who likewise had been born abroad, had come to Caltech during the 1930s, and who, like Malina, had worked closely with von Kármán: Louis Dunn, a native of South Africa, and William Pickering, a native of New Zealand. It was Pickering who led JPL's Explorer effort. After a four-year collaboration with von Braun and the Huntsville team, JPL's satellite was launched into space on the Huntsville group's Jupiter-C rocket.

In the years before and immediately after World War II, American universities and the US military continued to recruit Western-born scientists to bring the substance and style of European applied science to American aviation. Typically academics from upper- or upper-middle-class European families, these scientists made their mark not just in the fields of astronautical research, development, and exploration, but on engineering education. They "led the teams that led the way," to quote a Pickering biographer

(Mudgway)—and they established space exploration as a highly visible, powerful weapon in the Cold War.

In addition to those recruited for their scientific expertise, however, the Americans ultimately tapped ordinary citizens who sought to escape the German and Italian regimes, or who had sought earlier to escape persecution in the Soviet Union. Frank Toscelli was an Italian soldier imprisoned by the Germans in Italy in 1943. Like thousands of other Italian military, he escaped German evacuation to prison camps. Toscelli was lucky enough to happen upon a group of American soldiers in the course of his escape and sought shelter with them. His father had immigrated to the United States earlier, and Toscelli followed the Americans after the war, going to study in the United States. In 1953, he earned a BS in mechanical engineering at Carnegie Mellon University. "Sharing the experience of thousands of former Axis nationals eager to emigrate from war-torn Europe to the land of opportunity, Toscelli played a game of cat and mouse with US immigration officials as he tried to obtain an immigrant visa," Sylvia Doughty Fries writes (94). Neither an American bride nor a $1,000 bond posted by his uncle would do the trick. He finally got a visa by traveling to Cuba. He returned to work for Westinghouse Airbrake and later the company's Astronuclear Laboratory. He was recruited by NASA in 1964, earned his PhD in aerospace engineering, and went to work at Goddard Space Flight Center in Maryland, where he contributed as a senior engineer to "virtually every Goddard satellite program" (95).

Isaac "Ike" Petrovsky had come with his parents to Ellis Island in 1930 from Lithuania and became an engineer because it was the closest thing to a trade that he could study in college. In 1941, after the United States entered World War II, he entered ROTC. "I didn't have too many suits of clothing . . . They called us up and they said . . . 'you have to sign up now, because otherwise, you'll be drafted. This way you can become a second lieutenant.'" He spent the duration of the war servicing airplanes in British Guyana. Military service and engineering training had created bright opportunities for him; after working for a time in the private sector, he went to work for NASA at Kennedy Space Center in 1968 (Fries 10).

Like the immigrant engineers recruited to staff the very top of the American defense/academic/industrial complex, many of these more ordinary immigrants were the product of an early European education that placed a heavy emphasis on the liberal arts and sciences, a sensibility they brought to their highly technical work. However, these later European recruits tended to come from more humble backgrounds than their managers. The NASA engineers of Petrovsky's and Toscelli's generation "were the sons of an older industrial America, predominantly from the upper Midwest and the Northeast" (Fries 25). Reflecting general immigration trends to the United States in the 1930s, '40s, and '50s, they were a part of the European tide that formed the backbone of the country's older urban centers. A few were sons of engineers or highly educated people, but many came from working-class families in the trades and saw engineering as their ticket to economic mobility.

> Theirs was the class that experienced the pain of the Great Depression more than any other, with its longing for a more secure place in the world, for salaried middle-class stability. What they sought could be found most directly through the profession of engineering, unencumbered—partly because of its novelty—by the social barriers entrenched in some of the older professions, such as law and medicine. (25)

The postwar economy demanded labor. Not only immigrants, but their children too made their way into the American middle class by attending college in the United States and seeking employment in the aerospace industry. Some of these second-generation Americans made it to the uppermost level of the Apollo program, as the now aging European pioneers who had helped usher the space program through the initial years of the Cold War approached retirement. Rocco Petrone, who followed Peenemünde alumni von Braun and Eberhard Rees as the third director of the Marshall Space Center in Huntsville, was born not in Europe, but in Amsterdam, New York, of working-class Italian-born parents. He delivered ice to help support his family and won an appointment to West Point, where he played football. He went on to earn a master's in mechanical engineering from MIT, served in the Army, rising to Lieutenant Colonel, and ultimately worked on the construction and development of the Saturn launch rockets for the Apollo lunar missions. As director of Kennedy Space Center's launch operations from 1966 to 1969, Petrone became legendary for his tenacity, his deftness at political negotiations, and his attention to quality control. One NASA manager recalled a typical encounter with Petrone during this period:

> NASA was grilling the contractor people on some program delays. Rocco was the one who was never content with an answer and kept probing this one young contractor engineer, who quickly reached the limit of his knowledge. Instead of admitting it, he tried to bluff. Bad mistake. Rocco took him physically off of the podium and told the young man's boss to take the podium. He also told the boss that the young man was to be removed from the program. Brutal? Maybe, but it made us all know our subjects thoroughly from then on. (Bernstein)

It was his image as the hard-nosed son of working-class immigrants that made Petrone an attractive candidate to lead the Marshall Center, according to an official Marshall Center history. Taking the reins in 1973, Petrone earned a reputation as a cost-cutter in an era of constraint. As public support for the space program dwindled and the Nixon Administration demanded more austerity, reductions in force led to the demotions and forced retirements of many members of the original von Braun team. Petrone ended his career in 1975 as NASA's associate administrator, the third-ranking official of the agency. The reductions carried out at the Marshall Center illustrated not just a loosening of the grip of the older immigrant managers, but a concern with bringing in new recruits, including minority recruits (Dunar and Waring 167).

The immigrants and second-generation Americans who created the nation's space agency were part of a wave of pre- and immediately post-World War II European migration to the United States fueled mostly by oppression and economic hardship. Southern and Eastern Europeans were a major part of the nation's immigration mix prior to World War II. They remained so in the years immediately after the war, when their skills were desperately needed in the older industrial cities of the East and Midwest. For those immigrants and second-generation Americans who were educated or could gain an education through the GI Bill, the promise of economic mobility was great, and the defense industry offered a powerful push into the mainstream and the middle class (Dinnerstein et al. 249). As southern cities began to modernize at a rapid pace, these immigrants followed to the Sunbelt.

NASA and Post-1965 Immigration

The mix of immigrants working for NASA became increasingly diversified starting in the 1960s, reflecting changes in the nation's overall immigrant mix. The increasing desirability and prestige of post-secondary science education in the United States, advances in travel, and perhaps most importantly, changes in immigration law brought more foreign-born engineers and scientists from outside Europe. The Immigration and Nationality Act of 1965, also known as the Hart-Celler Act, eliminated the forty-year-old national origins quotas, which favored European immigrants and shut out people from Asia, Africa, and the Middle East. The Hart-Celler Act had the effect of opening America's doors to all nationalities on a roughly equal basis. Congress followed fifteen years later with the Refugee Act of 1980, aimed at creating an orderly planning process for accepting and settling refugees outside the limits permitted in the 1965 scheme.

Both of these laws had a profound impact on who could lawfully enter the United States, and ultimately what it meant to be American. At NASA, one early example of the coming shift was Farouk El-Baz, who retired recently as director of the Center for Remote Sensing at Boston University. El-Baz, born in Egypt, received his BS in chemistry and geology in Egypt, then traveled to the University of Missouri-Columbia for his master's degree. From 1967 to 1972, El-Baz was supervisor of Lunar Science Planning at Bellcomm, a company that conducted systems analysis for NASA headquarters in Washington, DC. In that position, he excelled at cataloging features on the landscape of the Moon from photos. He began teaching NASA astronauts how to observe the lunar surface from orbit and collect samples on the Moon to create a portrait of its surface. This scientific work earned El-Baz the roles of secretary of the Landing Site Selection Committee for the Apollo missions, principal investigator of visual observations and photography, and chairman of the Astronaut Training Group of the Apollo Photo Team.

Egyptian-born immigrants were relatively rare in the United States when El-Baz took on his leadership role with the Apollo program. Author Andrew Chaikin recalls Astronaut Ken Mattingly's response when he learned a scientist named Farouk El-Baz wanted to talk with him about the Moon. "Farouk El-Baz? Have we run out of geologists in this country?" he exclaimed (395). El-Baz himself persisted and became a one-of-a-kind mentor to the astronauts, who began to compete with one another to learn more about the geological composition of the Moon's surface. "It was work like hell, figure out everything yourself before you open your mouth, then, when you open your mouth, make sure that you're going to add something, because every single one of them is adding something," El-Baz said of his work with early Apollo teams. "You've got to add a lot more than they do so that you, as a weirdo, can be accepted amongst them" (Rebecca Wright).

Nowhere was the diversity of the space agency's immigrant workforce more evident than at NASA's JPL. JPL has continued to reflect the diversity not just of its immigrant talent but of its home community in Southern California. About 34 percent of Los Angeles County's population is foreign-born, again more than double the percentage in the country at large. By the 1970s, JPL's Latino and Asian American employees surpassed twice the proportion of these groups employed by NASA as a whole, reflecting continued high levels of Asian and Latin American immigration following passage of the 1965 Hart-Celler Act (Westwick). The laboratory's ninth director, Dr. Charles Elachi, appointed in 2001, was born in Lebanon to an ordinary middle-class family and grew up watching John Wayne films, often speaking with American tourists. As a child, Elachi was impressed by Americans' "open, positive attitude toward life, their attitude that anything is possible" (Elachi). Though he earned his undergraduate degrees in France, he had heard a great deal about JPL and Caltech, and he traveled to the United States for a master's degree and PhD in electrical sciences from Caltech, later earning a master's degree in geology from UCLA and an MBA from the University of Southern California. To Elachi, NASA's role is to perpetuate "the spirit of exploration." In a NASA oral history, Elachi recalled Kennedy's September 1962 speech at Rice University as a motivator: "Kennedy said it . . . we do these things because they are very hard to do and will expand our sphere of knowledge" (Rebecca Wright).

Kamal Oudrhiri, mission manager and Planetary Radar and Radio Sciences group supervisor at the Cold Atom Lab, JPL, has played a key role in the Cassini (Saturn), Grail (Moon), and multiple Mars missions, including the Mars Exploration Rovers (MER), "Spirit," "Opportunity," and "Curiosity." Oudrhiri grew up in the South of Morocco. From the time he first saw a photo of the Voyager spacecraft in a magazine, he dreamed of traveling to the United States and working for NASA. Though his family had concerns about his traveling to California, where he knew no one, he was relentless in pursuing his dream. After graduate studies at the University of Southern California, he joined NASA in 1996. "I really wanted to be part of it . . . I had this sense of discovery, exploration, even adventure," he told a podcaster. Though relatives had encouraged him to study in France, where he had relatives, he yearned for the challenge of a new life in a new land. "It was a personal drive . . . I lived an exploration myself," he said. Once established in the United States, he started a nonprofit organization, Grove of Hope, in Los Angeles that provides hands-on educational outreach to children in the US, Africa, and India with the hope of sparking a lifelong curiosity for careers in science and technology (Rafi, Hatim).

Like Elachi and Oudrhiri, most of the NASA immigrant scientists who immigrated post-1965 initially pursued higher education in the United States. Like Elachi's predecessor von Kármán, they have spurred students around the world to pursue graduate study in the United States. This sort of graduate work can make candidates eligible for special skills visas, a reliable ticket to the American Dream. And since the mid-1970s, the percentages of foreign students have more than quadrupled to nearly six percent of all university students enrolled. Their countries of origin have changed in keeping with the overall change in US immigration—while in von Kármán's day, they hailed primarily from Canada, Taiwan, India, and European and Latin American countries. Following passage of the 1965 Act, students from Asia have made up a larger share of international student enrollment (Israel and Batalova).

Reflecting both the breadth of American immigration post-1965 and the increasingly restrictive immigration regime of the 1980s and 1990s, NASA's foreign-born personnel have come to the United States in myriad ways, not just as students and young professionals. The 1965 Immigration Act had for the first time capped the number of visas available to those seeking to immigrate legally from Latin America. Bernardo Lopez, senior engineer at JPL, left his native, war-torn El Salvador right out of high school in the early 1980s and arrived in California as an undocumented immigrant. In 1976, gazing at the sky, he saw Comet West. "I thought it was so interesting," he said. "I remember reading in the paper that NASA scientists were observing the comet in Pasadena (California), and I thought I would like to do that kind of work, but I also knew it was very hard." He worked as a truck driver, went to community college, and later became a NASA engineer and a US citizen. Since 1998, he has worked on the design of eight spacecraft, most recently the European probe Mars Explorer (Denver Post).

Second-generation scientists of the new immigrant wave, like their second-generation predecessors, have told equally impressive stories of grit as they struggled with poverty and hardship. Marleen Martinez Sundgaard, engineer in charge of the testbed for the Mars InSight Mission, was born to a migrant farmworker family in Washington State and split her time growing up between American and Mexican schools. When she was eight years old, she won an essay contest that placed her in the Space Academy space camp in Huntsville, Alabama, where she "trained to be an astronaut for an entire week." She was hooked. "That was the beginning of me saying: I'm definitely not going to stay in the [agricultural] fields," she told an interviewer. "I'm definitely going into space. What do I need to do to get there?" Sundgaard studied aerospace engineering on scholarship at the University of Washington and has spent her career at the JPL in California (Zilm).

Unlike their pre-1965 predecessors, my unscientific sampling of NASA's post-1965 immigrants from the Middle East, Latin America, Africa, and Asia were engineers and scientists more likely to be members of ethnic, religious, and racial minorities in the United States. Like Oudrhiri, many express a concern with encouraging scientific education in those countries and among underserved populations in their adopted country. In turn, the newer immigrants offer an enormous reserve of soft power for the United States in the post–Cold War era. Dr. Kamlesh Lulla, director for University Research Collaboration and Partnership Office at Johnson Space Center in Houston, Texas, and called a "people to people ambassador for the United States," recalls listening to the Apollo 11 lunar landing on the radio as a young graduate student in India. When the US embassy in India exhibited Apollo artifacts around the country after the landing, he doggedly sought them out. "I stood in line, in one place, for about four hours just to get a glimpse of the spacesuit, the gloves—to make sure I didn't miss a chance to see all of that," he told an interviewer recently. Lulla applied to an international fellowship program, earned a second doctorate in the United States, and embarked on a more than thirty-year NASA career in Earth observations and remote sensing, mentoring many younger professionals in the process. Together with Astronaut Franklin Chang-Diaz, he forged a partnership with a Costa Rican university to train Costa Rican students to conduct remote-sensing activities and complete image analyses, culminating in the publication of a book of images of their country from space. With the first female shuttle astronaut, Sally Ride, he developed a program, EarthKAM, wherein schoolchildren may request images of specific landforms on Earth via the International Space Station. "Those are the kind of cultural values we grew up with," he said. "Working hard is great, but what are you doing for society? What are you doing for humanity?" (Williams).

Foreign-Born and Second-Generation Astronauts

Surprisingly, even the ranks of astronauts have included foreign-born Americans, as well as a number of second-generation Americans. Of the more than thirty Americans designated as astronauts since 1969, twelve were immigrants from countries including Argentina, Australia, India, and Peru (Robin Wright). A notable, if tragic, example is Kalpana Chawla, who was one of seven crew members who died on February 1, 2003, when the shuttle Columbia, after sustaining damage to its protective heat tiles during liftoff, became unstable during re-entry and broke apart over Texas and Louisiana (Haberman). Astronaut Chawla was born in Karnal, India. She immigrated to the United States in the 1980s and became the first Indian American to fly on a space shuttle crew. She traveled a total of roughly 6.5 million miles in space (Santora).

Other foreign-born astronauts include Piers Sellers, born in the United Kingdom. Astronaut Sellers received NASA's Distinguished Service medal, its highest honor, in 2016, and served as deputy director of its Sciences and Exploration Directorate. "When I was a kid, I watched the Apollo launches from across the ocean, and I thought NASA was the holy mountain," he is quoted as saying. "As soon as I could I came over here (to the US) to see if I could climb that mountain" (Carnegie). Sellers flew to the space station and on the space shuttle Atlantis in 2002 and 2010, and on Discovery in 2006, to perform research and carry out repair work. He made a total of six spacewalks over his career. Astronaut Franklin Chang-Diaz, Dr. Lulla's Costa Rican–born collaborator, flew seven missions aboard the space shuttle and developed a plasma rocket that "could blast the next generation of astronauts to Mars in just thirty-nine days." Chang-Diaz worked as a bank teller in his home country, saved enough money to go to the United States, and earned scholarships to attend the University of Connecticut and MIT. By 1979, Chang-Diaz had become a citizen, and the following year he became NASA's first Latino astronaut (Berger). Another Latinx astronaut is José Hernández, who started dreaming about space travel by watching coverage of Apollo 17 on television. Though he was born in the US, his parents were Mexican-born farmworkers, and he helped them seasonally, homeschooling himself during harvest months. Hernández earned a master's degree in electrical and computer engineering from UC Santa Barbara. After being rejected by NASA eleven times, he finally became an astronaut. In 2009, he participated in a fourteen-day mission to finish construction of the International Space Station. Now, Hernández runs his own foundation, Reaching for the Stars, which helps get youth in central California interested in STEM fields and provides first-generation high school seniors with scholarships (Blanco, Hernández, and Rubin).

Representing the Middle East is Iranian-American Jasmin Moghbeli, who was a Marine Corps major decorated for flying Cobra gunships on combat missions in Afghanistan. Moghbeli's family left Iran after the 1979 revolution and traveled to Germany, where she was born, before moving on to the United States. Originally Shiites, they converted to Lutheranism. As a sixth grader at Lenox Elementary School in North Baldwin, New York, she dreamed of becoming an astronaut and wrote a book report about Valentina Tereshkova, the Soviet cosmonaut who was the first woman in space. She studied aerospace engineering at MIT and enlisted with the Marines. In 2017, just as President Trump was announcing his Muslim ban, Moghbeli became an astronaut candidate—the first ever to have roots in the Middle East (Robin Wright). She is currently assigned to the Artemis Mission team at NASA and she may yet one day walk on the Moon (NASA.gov). In a 2017 interview for *The New Yorker*, Astronaut Moghbeli reflected on her newly earned title:

> I have family across the world . . . That just helps me remember how grateful we should be to live in this country. Yeah, it's got flaws, just like any other country. And there are things we can improve on, that we should absolutely work on improving. But, at the end of the day, we have amazing opportunities here. And the fact that I can be a female, Iranian, in the Marine Corps, and now becoming an astronaut—it's incredible. (Robin Wright)

Immigrant and Second-Generation Astro-Futurists and Space Investors

Immigrants and the children of immigrants not only flew and sent humans to outer space, but they also filled the ranks of astro-futurists who moved public opinion to support the idea of space exploration and travel. Wernher von Braun joined Willy Ley, a German transplant rocket historian, and Heinz Haber, a space/medicine expert, in producing a series of articles for *Collier's* magazine describing von Braun's vision of space exploration. The three went on to serve as advisers for a series of *Man in Space* television episodes that Disney produced in the 1950s. Von Braun had been inspired as a child by Jules Verne, and his futurist vision was brought to life on the TV screen by Walt Disney through animation (Mike Wright). A more recent figure, Carl Sagan, the American astronomer, cosmologist, astrophysicist, popularizer of science, and searcher for extraterrestrial intelligence, was the son of a Ukrainian immigrant garment worker. Writing during a period when human spaceflight was losing popular interest, Sagan worked to increase awareness of man's place in the universe, the realization that our planet is just one of many in space—surely influenced by the immigrant experience.

Immigrants also have pioneered as private investors, helping to build the space infrastructure of commercial and communications satellites and investing in space tourism. A number of private studies have documented the prevalence of immigrants among technology innovators. A 2016 study by the Information Technology and Innovation Foundation found that more than one-third of America's technology "innovators" are foreign-born (Nager et al.). More data is needed to get a better picture of how the foreign-born contribute to innovation, but anecdotal evidence is compelling. Peter Diamandis, founder of a constellation of space-related for-profit and nonprofit ventures including the X Prize Foundation and Space Adventures Ltd., a space tourism company, is the son of Greek immigrants. Diamandis early on expressed an interest in space exploration, later studying biology and physics at MIT and graduating from Harvard Medical School. His interest in space led him as a medical student to co-found the International Space University in Strasbourg, France, which grew from a summer program for high-achieving students into an accredited, world-class research institute (Caufield). Elon Musk is an immigrant from South Africa who attended Queens University in Canada and the University of Pennsylvania in Philadelphia. Musk is the founder of PayPal, SpaceX, and Tesla, and was behind the creation of the solar panel startup Solar City. Eren and Fatih Ozmen, Turkish immigrants and owners of the Sierra Nevada Corp., are currently at work on an unmanned cargo ship called Dream Chaser, designed to carry food, water, and science experiments to and from the International Space Station. Sierra Nevada joins competitors Orbital ATK and Musk's SpaceX in executing a contract worth up to $14 billion. The couple came to the United States separately as graduate students in the early 1980s, he for electrical engineering and she for an

MBA. They used their house as collateral in the purchase of Sierra Nevada, the small defense company where they both worked, for less than $5 million in 1994, and the company went on to earn $1.6 billion in 2017 sales, with 80 percent of its revenue from the US government (Debter).

Lessons from Scientists Without Borders

What can we learn from the stories of immigrants in the Apollo program and beyond? First, though dreams of space travel are not uniquely American, these dreams have been more easily realized in the United States. This, as we have seen, stems from a variety of reasons: the United States' historical desire to project power through its space program; cultural themes such as the "frontier," "get-to-work," and "immigrant" narratives that have supported space exploration; and finally the national economic means to support such programs.

Nearly a century after Theodore von Kármán labored to recruit a diverse corps of graduate students at Caltech's Guggenheim Aeronautical Laboratory, it is hard to avoid the conclusion that dreams of space travel are more easily realized in the United States *because of* the nation's mix of immigrants. As von Kármán suggested, their very diversity has provided a source of strength for such endeavors. As we continue to debate major changes to our nation's immigration system, we would be wise to take into account how those changes will affect the quality of our nation's science agenda, including its space program. Free movement of people and ideas is crucial to advanced science, as fifty-one Nobel laureates and tens of thousands of academics acknowledged when they signed a protest to President Trump's first immigration order, which would have instituted "extreme vetting" of refugees, commenced a temporary ban on allowing people from Iran, Iraq, Libya, Somalia, Sudan, Syria, and Yemen from entering the United States, and suspended the US Refugee Admissions Program for 120 days. Hundreds of students and scientists were affected by the later re-vamped Muslim ban (Yong).

Congress has been mulling a mix of immigration proposals over the past twenty years but has been unable to enact any major reforms. The Trump administration committed itself to reducing immigration overall by preventing illegal immigration, restricting protections for Dreamers, rolling back family-based immigration, cutting back on employment-based immigration, and cutting student visas. All this appears to have had the collateral effect of discouraging foreign students from studying in the United States. Overall graduate applications from prospective international students declined by four percent from Fall 2017 to Fall 2018, for the second year in a row, mostly fueled by decreased interest in US master's programs. Applications to graduate engineering programs, which make up about 25 percent of international graduate applications, fell 16 percent from Fall 2017 to Fall 2018, the second straight year of decline. International graduate applications in the physical and earth sciences declined by nine percent, and even in business, by eight percent. The greatest declines in enrollment came from the Middle East and North Africa, followed by Oceania and Europe; Chinese applications, by far the greatest percentage, were not affected. The greatest declines in master's applications were found at doctoral universities with the most research output. For individual universities, this has meant as much as a 30 percent drop in foreign enrollments from 2016 levels in some programs (Okahana and Zhou). Though the losses abated somewhat the following, pre-pandemic year, the COVID-19 pandemic again battered international graduate student enrollments in engineering, mathematics, and related disciplines (Zhou et al. 2020; Zhou and Gao 2021).

These figures suggest that the United States may be losing access to the best and the brightest of international students. Moreover, US citizen enrollment rates in engineering programs have also been dropping and the US continues to rank behind other countries in the percentage of students earning their first university degree in engineering or science. In several leading European and Asian countries, the percentage of undergraduates receiving their initial degree in engineering, computer science, physics and biological science, mathematics, and statistics is above 30 percent of all first-time degree recipients. In the United States, the corresponding figure is 22 percent (National Science Board 2018, 2020, 2022).

The shortage of qualified aerospace and aeronautics personnel is nothing new. For at least the past fifteen years, industry advocates and the nation's premier research scientists have consistently called for greater ease of immigration, along with stronger investments in US science education and domestic recruitment (American Institute of Aeronautics). "... (T)here has been a steady erosion in investment in the kind of scientific and engineering brainpower that keeps a nation competitive—and a consequent decline in American inventiveness," then-NASA Administrator Michael D. Griffin said in a 2006 speech to the National Society of Professional Engineers. "So anything that we can do to arrest these trends: to inspire young students to pursue technical careers and to motivate talented foreign-born graduate students to consider staying in the United States and work on the greatest exploration project of the twenty-first century—would be all to the good."

Griffin's words, now more than fifteen years old, still ring true. Americans continue to support space exploration: 72 percent of Americans say it is essential for the US to continue to be a world leader in space exploration, according to the Pew Research Center, and 80 percent of Americans surveyed say the International Space Station has been a good investment for the country (Fink and Strauss). In addition, most Americans favor increased levels of legal immigration: from 2001 to 2018, the share of Americans who favor increased legal immigration into the US rose twenty-two percentage points (from 10 percent to 32 percent), while the share who support a decrease has declined twenty-nine points (from 53 percent to 24 percent) (Doherty et al.). This support is surprisingly bipartisan in direction, if not intensity.

Speaking in 2010 at NASA's fiftieth-anniversary celebration, Griffin described what he saw as the agency's impact on American society. "NASA is the entity which captures what Americans believe are the quintessential American qualities. Boldness and the will to use it to press beyond today's limits. Leadership in great ventures." Add to those qualities grit and teamwork among people of all nationalities and a keen desire to transcend one's borders and explore new worlds. The space program is a validation of the American Dream. As a nation, we should consider very carefully changes in our immigration system—not just changes focused on student visas or employment-based immigration, but on family-based immigration, refugees, and temporary residents as well. As this study documents, Project Apollo and the space program were built by immigrants and their families from many nations. Altering the composition and amount of future immigration into this country will affect not only the quality of scientific work our nation accomplishes, but will adversely impact the S&E professions, important avenues for personal mobility, and participation in the American Dream for some of our most valuable new citizens.

Note

1. Many thanks to the NASA History Division, especially Steve Garber, who provided research materials and generously included me in the Apollo Dialogues Workshop, December 7, 2018.

Works Cited

American Institute of Aeronautics and Astronautics. *Recruiting, Retaining, and Developing a World-Class Aerospace Workforce: An AIAA Information Paper*, 6 Mar. 2013.

Anderson, Stuart. *Immigrant Founders and Key Personnel in America's 50 Top Venture-funded Companies.* NFAP Policy Brief, National Foundation for American Policy, December 2011.

Berger, Eric. "Immigrant's Dream Could Be Our Ticket to Red Planet." *Houston Chronicle*, 18 Apr. 2010, https://www.chron.com/news/nation-world/article/Immigrant-s-dream-could-be-our-ticket-to-Red-1597188.php.

Bernstein, Adam. "Rocco Petrone, 80." *Washington Post*, 31 Aug. 2006. www.washingtonpost.com, https://www.washingtonpost.com/archive/local/2006/08/31/rocco-petrone-80/286dab16-f0bf-4750-97c7-072aeefc4a78/.

Blanco, Octavio. "How This Son of Migrant Farm Workers Became an Astronaut." CNNMoney, 14 Mar. 2016, https://money.cnn.com/2016/03/14/news/economy/jose-hernandez-migrant-astronaut/index.html.

Brinkley, Douglas. *American Moonshot: John F. Kennedy and the Great Space Race.* HarperCollins, 2019.

Carnegie Corporation of New York. "Piers J. Sellers." 22 Mar. 2019, https://www.carnegie.org/awards/honoree/piers-j-sellers/.

Caulfield, Brian. "Peter Diamandis: Rocket Man." *Forbes*, 16 Jan. 2012, https://www.forbes.com/sites/briancaulfield/2012/01/26/peter-diamandis-rocket-man/.

Chaikin, Andrew. *A Man on the Moon: The Voyages of the Apollo Astronauts.* Penguin Books, 2007.

Debter, Lauren. "Meet the Unknown Immigrant Billionaire Betting Her Fortune To Take On Musk in Space." *Forbes*, 12 Feb. 2019, https://www.forbes.com/sites/forbesdigitalcovers/2018/07/11/meet-the-unknown-immigrant-billionaire-betting-her-fortune-to-take-on-musk-in-space/.

Denver Post Staff. "Immigrant Works Up to NASA." *The Denver Post*, 16 Dec. 2005, https://www.denverpost.com/2005/12/16/immigrant-works-up-to-nasa/.

Dinnerstein, Leonard, Roger L. Nichols, and David M. Reimers. *Natives and Strangers: A Multicultural History of Americans.* Oxford UP, 1996.

Doherty, Carroll, Jocelyn Kiley, and Bridget Johnson. "Shifting Public Views on Legal Immigration Into the U.S." *Pew Research Center for the People and the Press*, 28 June 2018, https://www.people-press.org/2018/06/28/shifting-public-views-on-legal-immigration-into-the-u-s/.

Dunar, Andrew, and Stephen Waring. *Power To Explore: A History of Marshall Space Flight Center 1960–1990.* NASA History Office, 1999.

Elachi, Charles. "The Path to Mars Goes from Lebanon to Pasadena." What It Means To Be American, A National Conversation Hosted by the Smithsonian and Arizona State University, 22 Sept. 2014, https://www.whatitmeanstobeamerican.org/journeys/the-path-to-mars-goes-from-lebanon-to-pasadena/.

Fink, Carey, and Mark Strauss. "Majority of Americans Believe It Is Essential That the US Remain a Global Leader in Space." Pew Research, 6 June 2018, https://www.

pewresearch.org/science/2018/06/06/majority-of-americans-believe-it-is-essential-that-the-u-s-remain-a-global-leader-in-space/.

Fries, Sylvia Doughty. *NASA Engineers and the Age of Apollo*. National Aeronautics and Space Administration Scientific and Technical Information Program, 1992.

Griffin, Michael D. "NASA at 50." *NASA's First 50 Years: Historical Perspectives*, edited by Steven J. Dick, Government Printing Office, 2010.

Griffin, Michael D. *The Next Generation of Engineers: Michael D. Griffin Administrator National Aeronautics and Space Administration Remarks at National Society of Professional Engineers Professional Development Conference Washington, District of Columbia January 19, 2006, in Leadership in Space*. National Aeronautics and Space Administration, 2008, https://www.nasa.gov/pdf/141294main_NSPE_19_Jan_06.pdf.

Haberman, Clyde. "Challenger, Columbia and the Nature of Calamity." *New York Times*, 20 Dec. 2017. NYTimes.com, https://www.nytimes.com/2014/06/02/us/challenger-columbia-and-the-nature-of-calamity.html.

Hanle, Paul. *Bringing Aerodynamics to America*. MIT Press, 1982.

Hatim, Yahia. "Moroccan Scientist Kamal Oudrhiri Wins NASA Exceptional Service Medal." *Morocco World News*, 1 Nov. 2019, https://www.moroccoworldnews.com/2019/11/285817/moroccan-scientist-kamal-oudrhiri-wins-nasa-exceptional-service-medal/.

Hernández, José M., and Monica Rojas Rubin. *Reaching for the Stars: The Inspiring Story of a Migrant Farmworker Turned Astronaut*. Center Street, 2012.

Israel, Emma, and Jeanne Batalova. "International Students in the United States," *Migration Information Source*. Migration Policy Institute, 14 Jan. 2021, https://www.migrationpolicy.org/article/international-students-united-states-2020.

Jacobsen, Annie. *Operation Paperclip: The Secret Intelligence Program That Brought Nazi Scientists to America*. 1st Edition, Little, Brown and Co., 2014.

Kennedy, John F. *A Nation of Immigrants*. Harper and Row, 1964.

Laney, Monique. "Operation Paperclip in Huntsville, Alabama." *Remembering the Space Age*, edited by Steven J. Dick. Government Printing Office, 2008.

Lundquist, Charles A. *Transplanted Rocket Pioneers*. University of Alabama in Huntsville, 2014.

Mudgway, Douglas. *William H. Pickering: America's Deep Space Pioneer*. National Aeronautics and Space Administration, 2008.

Nager, Adams, et al. *The Demographics of Innovation in the United States*. Information Technology and Innovation Foundation, February 2016, http://www2.itif.org/2016-demographics-of-innovation.pdf.

NASA.gov. *Astronaut Jasmin Moghbeli*. https://www.nasa.gov/astronauts/biographies/jasmin-moghbeli.

National Academy of Sciences. *Policy Implications of International Graduate Students and Postdoctoral Scholars in the United States*. National Academies Press, 2005.

National Aeronautics and Space Administration, *Report of the Advisory Committee on the Future of the U.S. Space Program*. US Government Printing Office, 1990.

National Science Board, National Science Foundation. *Science and Engineering Indicators 2022: Higher Education in Science and Engineering*. National Science Board. https://ncses.nsf.gov/pubs/nsb20223/.

National Science Board, National Science Foundation. *Science and Engineering Indicators 2020: The State of U.S. Science and Engineering*. National Science Board. https://ncses.nsf.gov/pubs/nsb20201/.

National Science Board, National Science Foundation. *Science & Engineering Indicators 2018 Digest*. National Center for Science and Engineering Statistics, January 2018. https://www.nsf.gov/statistics/2018/nsb20181/.

Neal, Valerie. *Space Flight in the Shuttle Era and Beyond: Redefining Humanity's Purpose in Space*. Yale UP, 2017.

Neufeld, Michael. *Von Braun: Dreamer of Space, Engineer of War*. Alfred A. Knopf, 2007.

Okahana, Hironao, and Enyu Zhou. *International Graduate Applications and Enrollment: Fall 2018*. Council of Graduate Schools, 2019.

Ono, Hiro. *Immigrants in NASA: The American History of Giving Dreams a Chance*. Hiro Ono's Odyssey, Blog, 29 Oct. 2016, https://hiroono.com/en/2016/10/29/immigrants-in-nasa/.

Piszkiewicz, Dennis. *The Nazi Rocketeers: Dreams of Space and Crimes of War*. Praeger, 1995.

Rafi, Boushra. "Exclusive Interview with Kamal Oudghiri, Moroccan in Charge of Monitoring NASA's Mission to Mars." *Morocco World News*, 18 Oct. 2012, https://www.moroccoworldnews.com/2012/10/61442/exclusive-interview-with-kamal-oudghiri-moroccan-in-charge-of-monitoring-nasas-mission-to-mars/.

Report of the Advisory Committee on the Future of the U.S. Space Program. December 1990, https://history.nasa.gov/augustine/racfup3.htm#Excellence.

Santora, Marc. "Mayor's Sentiments Were Right, If the Astronaut Wasn't." *New York Times*, 12 July 2004. NYTimes.com, https://www.nytimes.com/2004/07/12/nyregion/mayor-s-sentiments-were-right-if-the-astronaut-wasn-t.html.

United States. Cong. H. Committee on Science and Technology. *Decisions on the Future Direction and Funding for NASA: What Will They Mean for the U.S. Aerospace Workforce and Industrial Base?* Hearings 111th Cong., first sess. Washington: GPO, 2009.

United States. Cong. S. Subcommittee on Science, Technology, and Innovation of the Committee on Commerce. *2006 Nobel Laureates*. Hearings 110th Cong., first sess. Washington: GPO, 2007.

Von Kármán, Theodore, with Lee Edson. *The Wind and Beyond: Theodore von Kármán, Pioneer in Aviation and Pathfinder in Space*. Little, Brown and Co., 1967.

Walsh, James P. "The Impact of Foreign-born Scientists and Engineers on American Nanoscience Research." *Science and Public Policy*, vol. 42, no. 1, 107–20.

Westwick, Peter. "The Jet Propulsion Laboratory and Southern California." Societal Impact of Space Flight, edited by Steven A. Dick and Roger D. Launius. National Aeronautics and Space Administration, 2007.

Witte, James. "Immigrants to America, Alfred Nobel, Mark Zuckerberg and the 2015 Nobel Prizes." *Huffington Post*, 8 Dec. 2015, https://www.huffingtonpost.com/james-witte/immigrants-to-america-and_b_8720402.html.

Wright, Mike. "The Disney-Von Braun Collaboration and Its Influence on Space Exploration," Marshall Space Flight Center History Office, https://history.msfc.nasa.gov/vonbraun/disney_article.html.

Wright, Rebecca. *Interview with Farouk El-Baz.* NASA Johnson Space Center Oral History Project, Boston, Massachusetts, 2 Nov. 2009, https://historycollection.jsc.nasa.gov/JSCHistoryPortal/history/oral_histories/El-BazF/El-BazF_11-2-09.htm.

Wright, Rebecca. *Oral History, Charles Elachi.* NASA at 50 Oral History Project, Edited Oral History Transcript. Pasadena, California, 23 Apr. 2007, https://historycollection.jsc.nasa.gov/JSCHistoryPortal/history/oral_histories/NASA_HQ/NAF/ElachiC/ElachiC_4-23-07.htm.

Wright, Robin, "Jasmin Moghbeli, Badass Astronaut." *The New Yorker*, 2 July 2017, https://www.newyorker.com/news/news-desk/jasmin-moghbeli-americas-badass-immigrant-astronaut.

Yong, Ed. "Trump's Immigration Ban Is Already Harming American Science." *Atlantic*, 29 Jan. 2017.

Zhou, Enyu, Radomir Ray Mitic, Christian P.L. West, and Hironao Okahana. *International Graduate Applications and Enrollment: Fall 2019.* Council of Graduate Schools (CGS), 2020.

Zhou, Enyu, and Janet Gao. *Graduate Enrollment and Degrees: 2010 to 2020.* Council of Graduate Schools (CGS), 2021.

Zilm, Kerstin. "Dreaming of Mars, a Scientist Inspires Others Along the Way." KCRW Greater LA podcast, hosted by Steve Chiotakis, 12 June 2019, https://www.kcrw.com/news/shows/greater-la/potato-farmer-turned-nasa-engineer-wants-to-visit-mars/dreaming-of-mars-a-scientist-inspires-others-along-the-way.

6

Picturing Women in the Space Age

The Impact of the Lunar Landing on Films, Television, and Fashion

JULIE WOSK

In 2019, the year celebrating the fiftieth anniversary of the Apollo 11 lunar landing, an exciting breakthrough event in space history was scheduled. March 29, 2019, was supposed to mark the first all-female spacewalk. American Astronauts Anne C. McClain and Christina H. Koch were scheduled to do a spacewalk to replace some of the huge lithium-ion batteries for the solar panels used to help power the research laboratory in the International Space Station, 200 miles above the Earth.

However, NASA announced just a few days earlier that it had only one of two medium-sized spacesuits ready so it had to cancel the event. EVA (extravehicular activity) spacesuits are modular constructions with custom-fitted torsos, arms, and legs, and one of the two medium-sized torsos could not be properly reconfigured in time for the spacewalk. (Only Koch would go, accompanied by male astronaut Nick Hague.)[1] The planned milestone event revealed the exciting headway women were continuing to make in NASA's space program, but the spacesuit snafu was a reminder of the delays and roadblocks that women astronauts have had to overcome.

Spacesuits have always had a special place in our cultural imagination, and early on, they became associated with the male astronauts who wore them. In documentary films about America's space program, there is often that galvanizing moment when the men suit up. The suits—life-protecting, made of highly sophisticated materials and technologies—are almost like icons in themselves. The Smithsonian's National Air and Space Museum in Washington, DC, houses the protective polycarbonate and silicone pressure helmets worn by Astronauts Buzz Aldrin and Michael Collins for the Apollo 11 lunar landing in 1969, as well as Neil Armstrong's spacesuit that he wore on the Moon landing. (In July 2019, it was refurbished and put back on display to help celebrate Apollo 11's fiftieth anniversary.)

But American women have had to wait to join this select group of astronauts. In the 1960s and early '70s, when America was in the midst of the heady days of the Women's Movement, women had aspirations but were excluded from NASA's space program. During 1960–61, a group of thirteen American women—later called the Mercury Thirteen, including Geraldyn ("Jerrie") Cobb—participated in the "Woman in Space" program, directed by Dr. William Randolph Lovelace at his research center in Albuquerque, New Mexico, and they successfully passed the same physiological tests given to male astronauts. As Margaret Weitekamp has written, "the Albuquerque tests revealed that women could demonstrate high levels of physical fitness, emotional stability, and mental endurance" (186). But these women were never admitted to NASA's training program because they lacked the requirement that astronauts be graduates of military jet pilot testing programs and have engineering degrees. (Because American women were not yet allowed to be fighter pilots, they could not gain admission to the country's two military training schools offering test pilot instruction.)

Lovelace's program was scrapped when the Navy withdrew its cooperation, and NASA played a role in this withdrawal of cooperation (Weitekamp 126).[2] At a United States congressional hearing in 1962, probing whether there was discrimination in NASA's establishment of qualifications for the astronaut program, Ohio representative Walter Moeller of the US House of Representatives said, "If today our priority program is getting a man to the Moon, maybe we should ask the good ladies to be patient and let this thing get accomplished first and then go after training women astronauts" (United States 71).

In 1963, the Russian Cosmonaut Valentina Tereshkova became the first woman in space, but it wasn't until 1978 that the first group of American females was admitted into NASA's astronaut training program. These pioneering women astronauts finally had their own milestone when Dr. Sally Ride, in June 1983, became America's first woman in space during her six-day STS-7 space shuttle mission aboard the Challenger (fig 6.1). That year Ride not only made her ground-breaking flight but also got to join her male counterparts in the ritual celebrating her achievement, as she was

Figure 6.1. Dr. Sally K. Ride, America's first woman in space. Photo: NASA.

photographed happily donating her spacesuit to the National Air and Space Museum.

The achievements by women like Sally Ride made the earlier fantasies about space women and celestial women a reality. Since ancient times there has been a fascination with Moon goddesses and female celestial beings. In Greece she was called Selene or Artemis (Artemis was the twin sister of Apollo); in Rome she was Luna; in China, Chang'e. In the modern world, fashion designers, television producers, filmmakers, and advertisers created fanciful images of lunar ladies and spacewomen—images that were often shaped by gender stereotypes while also sometimes subverting these stereotypes.[3] Artists have also long been intrigued by Moon goddesses, including the present author, Julie Wosk, whose painting *Artemis*, inspired by an ancient Greek sculpture, suggests the goddess's delicate face and powerful strength (fig 6.2).

Though American women in the 1960s and early '70s faced obstacles to becoming astronauts, and thus had no space helmets of their own, advertisers and fashion designers used space motifs to give women the illusion that they could participate in the excitement of the space age, even if they could not lift off into the cosmos. An advertisement for a Frigidaire "Gemini 19" refrigerator in 1966 featured models wearing bubble helmets with space-age dress designs by John Kloss. (The refrigerator featured not only

Figure 6.2. Julie Wosk, *Artemis*, 1994, oil on canvas.

space-age fashions but also modern technology, advertised as "the revolutionary space age successor to the old fashion compressor.") A Smirnoff vodka advertisement pictured a spacesuit-clad woman with a towering bee-hive hairdo (Marie Antoinette would have admired this). Asked the ad puckishly, "Haven't tried Smirnoff? Where in the world have you been?" The ad appeared in *Playboy* in 1966, and the magazine's male readers must have enjoyed this "space girl" savoring her drink.[4]

What these images masked was that there was a big gap between the way "space women" were portrayed in fashion and the media and women's actual aviation capabilities. Women aviators had long shown their prowess as pilots, from the early days of flying bi-planes through the achievements of pioneering female pilots like Louise Thaden and Amelia Earhart, and later, the skilled Women Air Force Service Pilots (WASPs) serving to ferry aircraft from airplane factories during World War II. In an iconic photo from 1944, a group of WASP flyers are shown with their training plane, the B-17 Flying Fortress aircraft, "Pistol Packin' Mamma," at the four engine school at Lockbourne Field, Ohio. ("Pistol Packin' Mamma" was a popular song in 1943.) These female flyers often said they enjoyed aviation because it gave them a wonderful sense of freedom and release. They could, if only for a while, lift off from the many social restrictions and stereotyped views that often bound women's lives.

Not only were women excelling as pilots, but they were also engaging with NASA's space program in another capacity. In the early years of America's space program, there was a group of women who supported the program using their extraordinary abilities in math and early computer use. After World War II, African American women, including Katherine Coleman Goble Johnson, participated in NASA's space program as mathematicians and computer programmers, as so compellingly portrayed in the 2016 Hollywood film, *Hidden Figures*.

But in other ways, as I detailed in my book, *Women and the Machine*, women were often represented as being grounded in their more traditional roles. In early air meet posters, they were often pictured simply as decorative figures waving at male pilots flying overhead, and some early male flight trainers were skeptical about the abilities of women aviators (Wosk 155–59). In some cases, the skepticism continued for decades. As recently as 2014 when Martha McSally, the first American woman to fly in combat, was running for Congress, an Air Force Lieutenant said of her that, as a pilot, she showed "incredible ineptness in the air" and that back in 1965 she had a "severe lack of knowledge and credibility" (Day, Philipps, and Oppel A13).

In the '60s and '70s, when women had aspirations but were excluded from America's space program, they didn't get to experience the sense of exhilaration and freedom that came from participating in spaceflight. Instead, advertisers used space images to suggest a different kind of freedom. Tampons for menstrual periods were considered liberating for women, and an ad for TAMPAX touted women's freedom by showing a young woman wearing a space bubble helmet as she floated aloft in the sky.

Even girdle ads sometimes associated the constricting garments with the freedom of spaceflight. The Apollo 11 spacesuits were manufactured by ILC Dover, a subsidiary of the International Latex Corporation, the company better known for manufacturing Playtex women's bras and girdles. Both spacesuits and undergarments benefited from the introduction of lighter synthetic materials. As Matthew H. Hersch has written, "new techniques for the manufacture of civilian apparel infused pressure suit development" (13). As seen in their advertising illustrations, Playtex somewhat incongruously suggested that by wearing their girdles, women would have the freedom to leap and dance in the air.

Space-age advertising and fashion came together in 1965 when the Mary Wells ad agency hired famed Italian fashion designer Emilio Pucci to create Braniff Airways' new uniforms and outfits for its flight attendants (then called stewardesses). At the time, the uniforms of flight attendants were often tailored and formal (Pollock 110).[5] The Braniff ads, however, suggested that the space-inspired fashions would bring "The End of the Plain Plane," and that the attendants dressed in Pucci were competent, sexy, and chic.

These images of women in the Braniff ads point to a recurring aspect of how space-age women were portrayed. In dueling cultural representations, they appear as women having competence and agency, as well as being attractive objects of beauty and desire. There was a similar duality historically in images of female aviators who were celebrated for their achievements yet also counterpointed by images of women grounded—pretty females sitting on the ground wistfully watching male aviators flying overhead. Implicit in these images is the notion that women could soar to new heights yet also be mindful to not venture out too far from their more familiar gendered roles—as wives, mothers, or decorative creatures to adorn posters, advertising, and art (Wosk 2001, 158).

One of the most striking features of Pucci's Gemini IV line for Braniff was the attendants' clear plastic bubble helmets, called "bolas" by Pucci, worn to protect their tall bee-hive hairdos from the rain as they walked between terminal and plane. The Pucci helmets, echoing the look of actual space helmets, were made of acrylic in this era fascinated by the new plastics. But though they were fashionable, they were impractical and were discontinued: they cracked easily and were difficult to store on the plane, and covered Jetways at many airports made them unnecessary (Pollock 108, 111).

Pucci, who had been in the Italian Air Force before he became a fashion designer, gave flight a look of excitement and frisson with his series of brightly colored dresses and uniforms—pink, turquoise, purple—made of soft printed jersey, silk, and gabardine. Braniff added a note of sexuality in its ad titled "Air Strip," picturing flight attendants changing outfits. They went from a cold-weather coat and plastic helmet to an indoor raspberry suit, then a serving dress that Braniff called a "Puccino"; after dinner, they could slip into "way out" clothes (Pollock 112).[6] (The sexuality was not lost on punsters who talked about "Pucci Galore"—a waggish reference to the sexy character, Pussy Galore, in the popular James Bond film, *Goldfinger*, from 1964.)

In the years leading up to the lunar landing and after, the space age continued to generate excitement in the world of fashion. Even though women could not yet participate as astronauts, in the 1960s and 1970s fashion magazines and American and European designers, including André Courrèges, Pierre Cardin, and Paco Rabanne, created fashions for the galactic female who wanted the look of space-age chic. (For even more glamour, space-minded women could also wear lipstick with celestial names, like Clairol's "Moon Babies" lipstick in 1966.)

One of the most iconic images of the "space girl" during this period was Richard Avedon's photograph of the British supermodel, Jean Shrimpton, which appeared on the cover of *Harper's Bazaar* magazine in April 1965. Wearing a hot pink space helmet designed by the milliner Mr. John, Shrimpton in the photo gazes straight ahead, her one holographic blue eye blinking flirtatiously at the viewer. This is a space woman with agency—engaging, provocative, and self-assured.

The April *Harper's*, which celebrated the magazine's twentieth anniversary, included photographs of Shrimpton and model Naty Abascal wearing silver Project Mercury spacesuits on loan from NASA. The magazine also featured "Galactic Girls"—models wearing bubble headpieces and standing in a landscape simulating the Moon's Sea of Tranquility (actually, a Fort Lauderdale quarry) (Mears).[7] Unlike Shrimpton's cover model with her audacious eye, these space women are simply pretty creatures to be looked at. Standing in a mock lunar landscape, one model (said *Harper's*) was wearing a dress designed "to catch the eye of the astronauts," turning her into an attractive object of the male gaze.

French designers, too, captured the allure of the space age. Pierre Cardin created his "Cosmocorps" collection, and during Paris Fashion Week in 1968, he presented models dressed in sleek silver-colored vinyl (fig 6.3). After NASA Astronaut Alan Shepard made America's first spacewalk in 1961, Cardin talked to the astronaut, and on October 17, 1969, he toured NASA's Manned Spaceflight Center in Houston. He saw the need for improved spacesuits which he said "were fascinating" but too bulky (Bender 42). Cardin's interest in space continued: his "Cosmos day ensemble" featured a red minidress for women complete with a molded plastic visor on a helmet-like hat. In 1970, actress Raquel Welch was photographed wearing a Cardin blue vinyl miniskirt and headpiece with a Plexiglas visor.

But it was designer André Courrèges who was widely dubbed the "Space Age Couturier." In the spring of 1964, he introduced his "Space Age" collection with clothes embodying "The Moon Girl Look." Said Courrèges, who had served in the French Air Force and was trained as a civil engineer, "I love the American spirit. The spirit of going to the Moon" (*Women's Wear Daily*).

His white and silver fashions gave women a sense of freedom and mobility by featuring miniskirts, flat-soled white go-go boots, silver-sequined trousers, A-line or trapezoidal skirts, and dresses with cutouts. If they could not yet be astronauts, women wearing the designer's clothes could at least fantasize about an extraterrestrial frolic.

When photographer Annie Leibovitz was commissioned to create an ad for Louis Vuitton years later, the fantasy space-age females were replaced by a real-life female astronaut. In June 1961, Richard Avedon had photographed three astronauts, Alan Shepard, Gus Grissom, and John Glenn, wearing Mercury spacesuits for the June 1961 *Harper's*, but in 2009, Leibovitz created a photo for a Louis Vuitton luggage ad celebrating the Moon landing's fortieth anniversary that not only pictured Apollo 11's Buzz Aldrin and Apollo 13 Pilot Jim Lovell, but also Astronaut Sally Ride—all sitting on the hood of a pickup truck as they looked up at the night sky, evocatively illuminated by a full Moon (Kiefaber). (Oddly, the $1,530 Vuitton travel bag in the advertisement was named "Icare," an apparent reference to Icarus, the son of Daedalus in ancient Greek mythology who disobeyed his father, flew too near the sun with his waxen wings which melted, and fell into the ocean to his death. We are left wondering: was being Moon-struck and carrying the piece of Icare luggage a celebration of the cosmos or actually a tragic act of hubris?)

Early Films and Television

In the 1960s, with all the excitement about the space age and the impending Moon landing, space women were also featured in the world of television and films. There had been fantasies about life on the Moon in the early years of film history: space women appeared in French filmmaker Georges Méliès's 1902 short film *Le Voyage dans la Lune* (*A Trip to the*

Figure 6.3. Space-Age Fashion by Pierre Cardin. Photo by Keystone/ Hulton Archive/Getty Images.

Moon) and German director Fritz Lang's 1929 film *Frau im Mond* (*Woman in the Moon*), based on a novel and script both written by his wife, Thea von Harbou.

In Méliès's film, a group of men—a professor and a group of male astronomers—are the ones who use the technology, traveling to the Moon using a capsule or rocket shot from a cannon, while pretty, shapely women wearing skimpy outfits—sailor's middy blouses and exposed legs—appear in their conventional roles as attractive decoration. They stand guard by the capsule, and later, when the men are on the Moon, they are put on display through the use of special effects: their smiling faces appear in stars overhead, and the actress Bleuette Bernon as Phoebe sits picturesquely on a crescent Moon—an image popular in postcards of the period.[8] In a celebratory parade after the men return to Earth, the women, now carrying rifles, do some serious work as they tow the rocket, but soon they join in the dancing.[9]

Even before the Russian launch of Sputnik in 1957, which propelled America into a space race, the idea of interplanetary travel and living in space also had great appeal in early American television programming. During the 1950s, American viewers could watch "space operas" like the popular show *Captain Video and the Video Rangers* (1949–1955) and *Space Patrol*, which aired from 1950 to 1955.[10]

Television and later films also often presented dueling cultural images: space women who embodied space chic and space savvy—glamorous female characters who were scientifically and technically adept. The women in *Space Patrol* were very different from female characters that often appeared in American television sitcoms and appliance advertisements of the 1950s, who were pictured as happy homemakers clueless about all things mechanical, women working in their kitchens wearing high heels (Wosk 2001, 230).

Space Patrol's female characters were not just pretty window dressing but were often portrayed as brave, resourceful, and skilled pilots; unlike cultural stereotypes of the female baffled by all things mechanical, they were skilled in science and technology. The two central female characters in the series were Carol Carlisle (played by Virginia Hewitt) and Tonga (played by Nina Bara), both of whom could pilot their spacecrafts.[11] Tonga, who at first was the villainous, evil "Lady of Diamonds," was transformed into the trusted Chief of Security for the United Planets Space Patrol, which was charged with keeping planetary peace. (Bara also played the evil Alpha in *Missile to the Moon*—a 1958 film with a huge menacing spider and a Miss Universe pageant on the Moon.) Carol in *Space Patrol* was a skilled scientist with a degree in astrophysics. In the 1951 episode, "The Agra Ray," she tests her invention, which speeds up plant growth to help eliminate food shortages but is also a dangerous weapon that in the wrong hands could turn a city to stone. In "The Disastrous Flight to Pluto," Tonga and Carol also work on an atomic device that threatens to endanger their space yacht with radiation. Said Hewitt about *Space Patrol*: after the trauma of the Second World War, the series provided "pure escapism, fantasy." She added, "Who would have dreamed, in those days, that in a few years there'd be Sputnik, and then somebody *really* up in the Moon?" (Bassior 311).

In America during the 1960s, amidst the excitement about the space race and the impending Moon landing, space women were again featured in the world of television and films. In 1961, America's space program was expanding, propelled by early Russian successes in space. On April 12, 1961,

Cosmonaut Yuri Gagarin became the first human in space. One month later, American Astronaut Alan Shepard became America's first man in space, and in 1962, John Glenn became the first American astronaut to orbit the Earth.

In American sci-fi television series of the sixties, space-age women revealed changing views of women's roles: they could still be portrayed as housewives and sexy sirens but also as sassy females who defied gender stereotypes. The animated American television series *The Jetsons* (which first aired in 1962–63 and then again in 1985–87) spoofed space-age fantasies about a future where the Jetson family traveled in flying cars and lived in "Spacepad" apartments furnished in a type of space moderne style. The show comically presented two gender paradigms: the mother, Jane Jetson, is a stereotypical housewife who loves shopping and using her space-age push-button appliances, while her maid, Rosie the Robot, is a comically outspoken, smart, no-nonsense female who is a whiz at helping the children with their math and homework (Wosk 2015, 126–28). Spoofing space-age fashions, in *The Jetsons* episode "Miss Solar System" (which aired February 3, 1963), Jane wants to enter the Miss Solar System Pageant and goes shopping at the "Satellite Dress Shop," where she buys an original dress in the "Martian Fall Line" designed by Pierre Martian—a comic reference to the space-age fashions of Pierre Cardin.[12]

In this decade of the Women's Movement, the Civil Rights Movement, and the space race, African American actress Nichelle Nichols played the role of Lt. Uhura, the female communications officer onboard the Enterprise, in the American television series *Star Trek* that first aired from 1966 to 1969 (fig 6.4). As the ship's communications officer, her role was serious, but she was dressed in the popular style of the era: miniskirt and go-go boots. While women at the time were wearing girdles that Playtex claimed gave them the feeling of freedom, Nichols pointed out that her character's name, Uhura, came from the Swahili word, "uhuru," meaning, "freedom." The name was apt since her character in the series pushed the boundaries of women's roles.

The first African American woman in space, Dr. Mae C. Jemison, who in 1992 flew as a mission specialist on space shuttle *Endeavour,* noted that as a young girl, she was inspired by the *Star Trek* series and in particular, by Lt. Uhura. Jemison herself played the role of Lt. Palmer in the 1993 television series *Star Trek: The Next Generation,* in the sixth season episode, "Second Chances" (fig 6.5).

Figure 6.4. Actress Nichelle Nichols played the role of Lt. Uhura in the American television series *Star Trek*.

Figure 6.5. Dr. Mae C. Jemison, American engineer, physician, and NASA astronaut. Photo: NASA

Figure 6.6. Actress Dee Hartford as the female android named Verda in the 1966 American television series, *Lost in Space*.

Space women in 1960s American television also appeared as sexy androids that upended gender stereotypes. In the 1966 television series, *Lost in Space*, actress Dee Hartford in the sardonic episode, "The Android Machine," played the beautiful android, Verda, who is accidentally ordered by Dr. Smith from a vending machine (fig 6.6). In this episode, the Robinson family and others who are castaways on a galactic planet are great admirers of Verda, who is silver-skinned and sports a futuristic headpiece. Aside from her bizarre headgear, Verda is smart, courageous, and savvy—a resourceful, compassionate creature who saves people on the planet with her mechanical know-how and self-sacrificing ways (Wosk 2015, 110–111).

Space Women in Films

In American films of the 1950s, as in American television, space women could be both sexy and fierce. In 1958, one year after the launch of Sputnik and the same year as the movie *Missile to the Moon*, director Edward Bernds produced his campy B movie, *Queen of Outer Space*. A group of male astronauts accidentally find themselves on the planet Venus, which is peopled by beautiful, sexy women carrying guns and wearing miniskirts and high heels (and conveniently speaking English). All the men, including male scientists and mathematicians, have been imprisoned. In its review of the film, *Variety* magazine, referring to an American burlesque show, mused that "Most of the female characters in *Queen of Outer Space* look like they would be more at home on a Minsky runway than the Cape Canaveral launching pad" ("Queen of Outer Space"). Their queen, Yllana, wears a menacing mask, but the planet's resident scientist is the sexy and sympathetic Talleah, played by actress Zsa Zsa Gabor, who wears a long, slinky, filmy dress and has her own lab complete with purple glass vials.

In the ancient Greek comedy *Lysistrata* by Aristophanes, the women reluctantly go on a sex strike to bring about peace in their warring city, but these 1950s space women in *Queen of Outer Space* have been almost brainwashed into thinking they are better off without men. Some of the men, meanwhile, comically utter the sexist views of the 1950s. Says one of the astronauts about a Venus woman, "How'd you like to drag that to the senior prom?" In a 1950s B movie like this, it is often hard to tell if what the men say is mockery or a mirroring of gender stereotypes of the period. Yllana's prime weapon is the cardboard-looking "Beta Disintegrator" intended to destroy the Earth; and one of the men remarks, "How could a bunch of women invent a gizmo like that?" He even wonders how women could control the gizmo's dangerous rays: "How could they aim it? You know how women drivers are." Says the group's womanizer, Lt. Larry Turner, "How could a doll as cute as that be such a pain in the neck?" The Venus ladies, says one of the men, are like all women he knows: "millions of miles from Earth and the little dolls are just the same."

The film, which some film critics cheerfully derided as "schlock" and a "campfest," has a happy ending in which some of the sex-starved women fall in love with the men and eventually revolt, but as in other sci-fi and horror films of the period, there are some more serious hints of cold-war anxieties: Yllana's face underneath the mask has been disfigured by atomic

Figure 6.7. Hollywood actress Jane Fonda in the 1968 film *Barbarella*. Photo: Paramount Pictures/Photofest.

radiation burns and there are underlying fears of a rogue state destroying the Earth.

A decade later, other films showed signs of cultural ambivalence in portraying futuristic females. In Roger Vadim's sardonic 1968 film, *Barbarella,* based on the comic book, the title character (played by Jane Fonda, who at the time was married to the director) has the freedom to be sexy but is also competent when recruited for a very dangerous mission—and all the while wondrously dressed in high-fashion space-age costuming (fig 6.7).

During the film's opening credits, she is clearly established as a sex object. While Braniff Airways in its advertising coyly called the flight attendants' clothing change an "air strip," there is no question about Barbarella's sexuality when, in the film's beginning sequences, she floats weightlessly in her space vehicle's cabin and does a striptease as she slowly peels off her metallic, armor-like spacesuit, one section at a time. First she uses her manicured fingers to pull off her gloves and leggings, revealing her bare legs, and soon she is completely nude, her long blonde hair waving in the air.

The film also had some witty fun with its scenic design. In the ship's cabin, there is a folding screen decorated with an image of French artist Georges Seurat's nineteenth-century Postimpressionist painting, *Sunday Afternoon on the Island of La Grande Jatte,* depicting fashionable Parisian women promenading in their stiff, long dresses with large, rounded bustles. The artificial look of these women confined in their Sunday clothes offers a sly contrast to Barbarella's free-wheeling sexuality and artlessness. The sexy space siren, though, has a serious task: Earth's president recruits her to catch the wayward Duran Duran, who has vanished with his invention, a "positronic ray" weapon that threatens war on the planet. The president tells her that she's the right person for the job because she's a "five-star astro-navigatrix"—a designation meant to establish her bona fides as a skilled space woman, though her sexy clothing tells another story.

With the space age's fascination with fanciful galactic fashions, Barbarella often changes outfits, all created by French designer Jacques Fonteray, except for the last, bright-green outfit designed by Paco Rabanne. (Rabanne was often mistakenly credited for all the film's costumes, even among fashion writers) (Lunden 185–211). Even though she's billed as a high-ranking "astro-navigatrix," it is her comic ability to survive the "Orgasmatron" and major obstacles while wearing her constant change of clothes that gives her whatever heft she has in this pop culture view of a female in space.

As women entered America's astronaut training program in 1978, there were signs of change in the 1970s depictions of space women. They still could be space sirens like Barbarella, but in Ridley Scott's 1979 film, *Alien*, second mate and warrant officer Ellen Ripley (played by actress Sigourney Weaver), as the *Nostromo*'s intrepid lone survivor at the end of the film, has a formidable role as she puts on a white spacesuit and successfully conquers the film's grotesque alien creature that gets ejected from the huge towing vehicle and destroyed. Putting on the spacesuit is emblematic as she becomes even more empowered, using her wits and computer skills to destroy the horrifically menacing creature.

The most compelling space survivor in later American films was undoubtedly mission specialist Dr. Ryan Stone (played by Sandra Bullock) in Mexican director Alfonso Cuarón's 2013 film, *Gravity* (fig 6.8). On her first space mission, Dr. Stone (who went into space with a NASA crew to

Figure 6.8. Sandra Bullock as the mission specialist Dr. Ryan Stone in Mexican director Alfonso Cuarón's 2013 film *Gravity*. Photo: Warner Bros./Photofest

help upgrade the hardware on the Hubble Telescope) uses her expertise and endurance to survive catastrophic damage to their space shuttle STS 157, the *Explorer.* She reaches an international space station, which itself is destroyed, and finally uses her skill, resourcefulness, and courage to return safely to Earth.

In the film, Stone says she likes the silence of space and jokes with Captain Kowalski (George Clooney) as they work to repair the Hubble. Soon, though, their mission is suddenly aborted by the threat of flying space debris from a destroyed Russian satellite, and they are forced to return to the *Explorer*. Cast adrift and tangled in her tethering, Stone is terrified but finally helped by the self-sacrificing Kowalski and she finally manages, as the mission's lone survivor, to get to a Chinese space station, *Tiangong*.

Cuarón's film brings a welcome layering to the paradigm of the formidable woman in space, bringing us close to Stone's moments of angst. Her space journey is not only one of endurance (she survives a fire, lack of oxygen, and mechanical failures) but also an interior journey, as she confronts the loss of her beloved daughter, her isolation, and her fatalistic feeling that she is going to die, as she says mournfully to herself, "No one will mourn for me. No one will pray for my soul."

Ultimately, Stone rallies as she puts her space helmet back on, uses a fire extinguisher to propel herself out, and survives a shattering ride back into Earth's atmosphere. Splashing into the water, she almost drowns but manages to take off her clothes and swim to shore. Stripping herself of her clothes, she is down to her essential self. This is not Braniff Airway's provocative "air strip" but a way to emerge as a new woman—a space woman determined to be alive as she takes her first tentative steps on land. Rather than being sexily costumed as in *Barbarella* or defined by her uniform as an astronaut, Stone shorn of all clothing emerges as a woman who has survived through her extraordinary resourcefulness, endurance, fierce courage, and skill.

A more problematic, almost surreal image of the space woman appeared in the very intriguing 2014 short film *Afronauts* by Ghanaian filmmaker and screenwriter Nuotama Frances Bodomo. The film presents an alternative cinematic retelling of a quixotic, real-life attempt in 1960s Zambia to achieve a lunar landing ahead of America and Russia. Edward Makuka Nkoloso, a grade school teacher who was director of Zambia's National Academy of Science, Space Research, and Philosophy, recruited 12 teenagers and trained them, including a 16-year-old young woman, Matha Mwamba. (The program never materialized, and Matha left after she became pregnant.)

In the film, the teenaged Matha Mwamba (played by the albino African American actress Diandra Forrest) is being groomed by members of the Zambian Space Academy in a rugged terrain to become the first person to land on the Moon.[13] Her training includes learning buoyancy and balance by rolling down a hill in an oil drum and bouncing from a tarp being held by the men. In a dream, Matha pictures herself in a spacesuit and space helmet, walking on the surface of the Moon, but there are suggestions of her trepidation. Forrest's pale, anxious face is sometimes a poignant counterpoint to the optimistic, expectant faces of her family and other members of the Zambian space agency.

Matha's aunt is skeptical and says, "Stubborn girl," adding, "Think they care about you? They just blow you up and turn you into fireworks." The film itself is partly a gloss on the pitfalls of colonialism (Northern Rhodesia had just become Zambia in 1964). Says one of the academy men, imagining Matha's arrival on the Moon, "I see them welcoming you," but, he warns her, "Do not impose Christianity on them, Matha. Do not impose the nation-states on them."

As in a dream, Bodomo's film has a fascinating blend of realism and the surreal. Donning her silver spacesuit, Matha gets shot up into space from the rag-tag Bantu rocket, as one of the men says, "Let us rejoice! My little space girl has gone to the Moon!" In this film—which intersplices the voice of newscaster Walter Cronkite narrating the actual July 1969 Apollo 11 Moon landing, and a clip of Neil Armstrong going down the spacecraft stairs—Matha, in a final frame, looks out with her shimmering, darting eyes, and then closes them during her "real"—or is it ambiguously just imagined?—arrival on the Moon. Bodomo's evocative film leaves us wondering yet hopeful that this fantasy might also be real.

Light years away from the campy and sexy female astronauts and planetary space sirens in B movies like *Queen of Outer Space* during the 1950s and 1960s, the female astronaut in *Proxima* (2019), by French writer and director Alice Winocour, probes the physical, psychological, and emotional stresses of being a female astronaut, one who must not only contend with skepticism about her skills but also do the delicate balancing act between being a caring mother and a woman engaging in her career.

In *Proxima*, the astronaut Sarah Loreau (Eva Green) is training at the cosmonaut center, Star City, near Moscow, to go on a year-long mission to the International Space Station and has to deal with her mission commander, Mike Shannon, an astronaut who is skeptical of her technical abilities as a woman. Even more difficult and painful for her, she is torn by having to leave her young eight-year-old daughter, Stella, who fears that her mother will die before she does. Saying goodbye to Stella behind the glass that separates them, Stella reminds her that she had promised they could see the rocket together.

Capturing the pain and poignancy of their impending separation, Sarah honors her promise to Stella, and the night before liftoff, she breaks quarantine and takes her daughter out on the launch field as dawn breaks so they can see the rocket together (she disinfects later in the shower). At liftoff, as mission control says, "Umbilical separation" and "All systems go," Stella has the reassurance she needed and smiles and claps as the spacecraft successfully launches, even as it takes her mother on a long journey so far, far away. As the film's credits roll at the end to the strains of Agar Agar's song "You're High," Winocour includes still photos of real-life women astronauts in uniform happily embracing their children. Through this reminder of reality, the director movingly illuminates the emotional complexities experienced by women who must try to reconcile these competing roles of mother and astronaut (fig 6.9).[14]

Figure 6.9. Eva Green and Zélie Boulant in *Proxima*.
Photo: Pathé/Photofest.

Later Space-Inspired Fashions, Spacesuits, and Groundbreaking Missions

Not only filmmakers but also twenty-first-century fashion designers continued to be fascinated by space imagery. Pierre Cardin's dress designs of the 1990s and 2000s evoked stars and galaxies with shimmering rhinestones, sequins, crystals, mirrors, and metallic surfaces, seen in his evening dress with orbital sleeves (1994, 2000) and his black velvet cocktail dress with parabolic sleeves (2016).

In 2017, Karl Lagerfeld added a mock rocket launch and fashions with astronaut and planetary prints to his Chanel Fashion Week show in Paris. The models wore quilted, metallic silver "space blankets," and with the push of a button, the rockets in a mock space launch emitted a boom, smoke, and fire from the engines accompanied by the sound of Elton John's song, "Rocket Man."

Creating a more ethereal air-to-space design, Japanese designer Maiko Takeda presented space headdresses in her 2013 millinery collection, Atmospheric Reentry (fig 6.10). The spiky headpieces—made of transparent plastic (Perspex) tinted with color gradients at the bases and tips, with

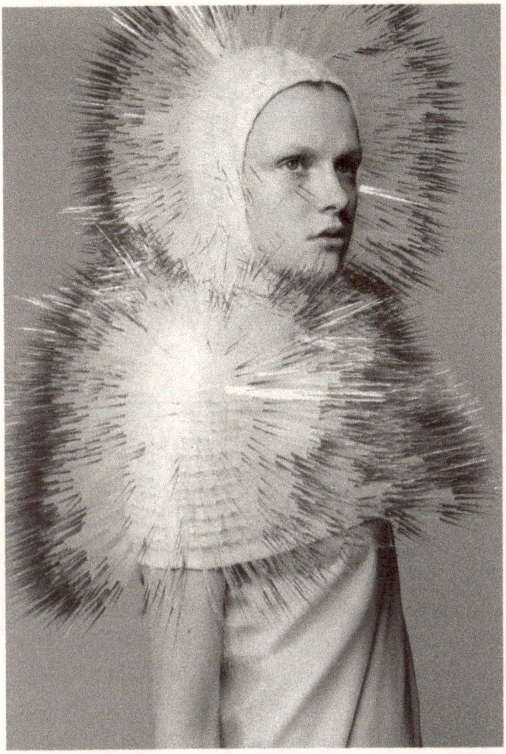

Figure 6.10. A space headpiece from Japanese designer Maiko Takeda's Atmospheric Reentry collection. © Maiko Takeda, Photo: Bryan Huynh.

small silver rings holding the spikes in place—were inspired by her seeing Philip Glass and Robert Wilson's 1976 opera, *Einstein on the Beach*. Takeda said of the opera, "Its futurist mood of the space age heavily influenced the aesthetic of my collection."[15]

Takeda's space-inspired fashions are evocatively ethereal and phantasmagorical, but in the real world of women astronauts and space engineers, important changes in actual spacesuit designs are underway. With increasing numbers of women enrolling to become astro-engineers and larger numbers of women in NASA's astronaut program, designers are turning their attention not to high fashion but instead to designing functional spacesuits with both women and men in mind.

Dr. Dava Newman, professor in MIT's Department of Aeronautics and Astronautics and former deputy administrator at NASA, is working collaboratively to develop new form-fitting EVA (extravehicular activity) pressurized MIT BioSuits, which she calls a "revolutionary approach" to new spacesuit design. The suits will be close to the skin, give the users greater

Fig. 6.11. MIT Professor Dava Newman wearing her BioSuit. (Dava Newman, MIT: Inventor, Science and Engineering; Guillermo Trotti, A.I.A., Trotti and Associates, Inc.: Design; Dainese: Suit Fabrication; Douglas Sonders: photograph.)

mobility, and will be shaped to the different bodies of men and women (fig 6.11).

Said Newman, "The BioSuit is a custom fit and custom design, so you can tell the difference between female, male, tall, short, etc. It's a unique suit for each unique wearer." The form-fitting suits will not only help the crew to identify their crewmates, but the suits for females, she noted, will have a special meaning: "The inspiration for girls is important," too, "because I do believe they need to 'see' themselves as astronauts and aerospace engineers to open up their minds and to allow themselves to accomplish these dreams!" (Newman).

Friday, October 18, 2019, marked an exciting milestone in American space history. Two American female astronauts, Ms. Christina Koch and Dr. Jessica Meir, became the first to engage in an all-female spacewalk (fig 6.12). For seven hours and seventeen minutes, 250 miles above the Earth,

Figure 6.12. NASA Astronauts Jessica Meir (*L*) and Christina Hammock Koch (*R*), who made the milestone achievement of completing the first all-female spacewalk. Photo: NASA.

they worked together replacing the International Space Station's controller, which "regulates the charge to the batteries that distribute solar power to the station" (Zraick A11).

In 1965 Russian Cosmonaut Svetlana Yevgenyevna Savitskaya became the first woman to walk in space, and in 1984 Kathryn D. Sullivan was the first American woman to do so—but Koch and Meir were the first to engage in an all-female spacewalk. During the live broadcast of the walk on television, listeners back on Earth also heard the calm, authoritative voice of another woman—NASA Astronaut Stephanie Wilson, who was the mission controller back at the Johnson Space Center in Houston—giving feedback and technical instructions to the women. She guided them to do critical tasks as they removed two bolts so they could exchange batteries and opened the airlock on the spacecraft so they could store the old battery. (She also heeded the astronauts' own suggestions, and said to one approvingly, "We like your plan!")[16] As Wilson also noted, "Spacewalks are one of the most dangerous things astronauts do," and she gave them friendly reminders to use the handrails of the spacecraft.

During the spacewalk there were reminders that these women were also human beings experiencing the excitement of their mission. When the spacecraft passed over California, one of them said, "Hello everybody in San Diego!" They were also modest: when President Trump on a telephone call congratulated them on their bravery (mistakenly saying they were the first women to make a spacewalk outside a space station), Meir quietly corrected him and gave credit to the many women who preceded them—female explorers, scientists, engineers, and astronauts. "This is just doing our job," she said, one for which they had been training for six years. She dedicated their walk not just to women but to "everybody that has a dream."[17]

The all-female spacewalk was an important milestone, and Christina Koch set her own record for the longest spaceflight in history by a woman after she spent 328 days in space, returning to Earth on February 6, 2020. More milestones were being planned—NASA was working on Artemis III, a plan to land the first female and the first person of color on the moon in 2025. And in 2019 NASA also unveiled its xEMU, a prototype of its next-generation spacesuit being designed for the Artemis trip—a spacesuit allowing for improved mobility and less discomfort, a suit that could accommodate differing body shapes, sizes, and genders. In March 2023, NASA presented a new prototype for the Artemis III spacesuit designs which would be manufactured by the company Axiom Space and feature greater flexibility, variable sizing, and adjustability, allowing the suits to be tailored to individual wearers.

Ultimately, from the ancient Greek goddess Artemis to the fashions, television, and films inspired by Apollo 11 to the female astronaut who will train for the future Artemis III mission to land on the Moon, images of women in space will continue to embody our cultural hopes, fantasies, and dreams.

Notes

1. McClain had already gone on a spacewalk earlier, so Koch was chosen to go using the one medium-sized spacesuit available.

2. The two training schools were the Naval Test Pilot School and the Aerospace Research Pilot School at the Edwards Air Force Base. (See also Lathers 18.)

3. Many such images were on display in the exhibit by curator Julie Wosk, "Imaging Women in the Space Age" at the New York Hall of Science, July 13, 2019–March 29, 2020.

4. Author's note: A few years after this ad appeared, I was hired to join the advertising department at *Playboy*'s corporate headquarters in Chicago, a far cry from my days as a Harvard graduate student the year before. Women weren't yet allowed in the editorial department but instead got to work at promotion and help create magazine ads like these.

5. Early flight attendants were nurses, and later attendants often dressed in a severe military style.

6. The campaign was successful: The number of Braniff Airways passengers was said to rise by 18 percent.

7. The images in the *Harper's Bazaar* issue were not just women wearing space gear. They also included a photo of the young Paul McCartney wearing a NASA suit, as well as photos of Ringo Starr and Bob Dylan.

8. The film was based on novels by Jules Verne and H. G. Wells, as well as Offenbach's operetta *Le voyage dans la lune*, and was popular internationally, including in the United States. As Matthew Solomon has written, Méliès had a "penchant for using the female body as special effect," and in the film's remake by Segundo de Chomón, women appear as dancing Moon maidens; and one is even brought back to Earth and gets betrothed to one of the astronomer astronauts (2, 10).

9. Commenting on the women, Victoria Duckett has provocatively argued that Méliès's film "parodies their irrelevance," and that the women put on exaggerated display "might actually be parodying the very performances they are setting in place" (163–4).

10. The show was aired first as fifteen-minute segments broadcast from Los Angeles and then thirty-minute shows broadcast from Los Angeles to New York (in a very early use of cables).

11. In another American television series of the time, *Tom Corbett, Space Cadet*, Dr. Joan Dale in an episode from November 1951, becomes the first woman to take command and gives cadets a tour of the Moon base.

12. A clip of *The Jetsons* episode was featured in the exhibit "Pierre Cardin: Future Fashion" at the Brooklyn Museum, July 20, 2019–January 5, 2020.

13. The landscape was actually a quarry in New Jersey.

14. *Proxima* was filmed in part at the actual Star City and lets us see the rigors of training with a centrifuge to withstand g-forces and being in an emergency situation underwater. Another 2019 film *Lucy in the Sky*, which was widely panned by audiences and film critics, starred Natalie Portman as the troubled astronaut Lucy Cola. The film was loosely based on the real-life case of American astronaut Lisa Nowak, an engineer, NASA astronaut, and a mission specialist in robotics, who drove cross-country with the intent of harming her lover's new love interest, US Air Force Captain Colleen Shipman, and was arrested for attempted kidnapping and other crimes.

15. In Act. IV, scene 3, a man goes up in a tube with a Moon-like shape (actually a compass).

16. The two astronauts also had other tasks, including tying down some of the insulation on a robotic arm.

17. In the film *Gravity*, Astronaut Ryan Stone confronts the dangers of becoming untethered, requiring the use of a jet pack. Astronaut Anne McClain, one of the commentators of the all-female spacewalk, reminded listeners that a similar situation could happen in the real-life drama of the spacewalk but fortunately, astronauts had safety equipment so that if they ever got untethered, they would have small jetpacks to get back to the space station. Still, said McClain, "I think they're having a lot of fun out there."

Works Cited

Bassior, Jean-Noel. *Space Patrol: Missions of Daring in the Name of Early Television.* McFarland, 2005.

Bender, Marylin. "Maxi? To Cardin, C'est Bon." *New York Times,* 14 Oct. 1969, p. 42.

Day, Helene Cooper, Dave Philipps, and Richard A. Oppel Jr. "'I, Too Was a Survivor': Senator McSally Ends Years of Silence." *New York Times,* 26 Mar. 2019, pp. A1, A13.

de Monchaux, Nicholas. *Spacesuit: Fashioning Apollo.* MIT Press, 2011.

Duckett, Victoria. "The Stars Might Be Smiling: A Feminist Forage Into an Early Film." *Fantastic Voyages of the Cinematic Imagination: Georges Méliès's Trip to the Moon,* edited by Matthew Solomon, SUNY Press, 2011, pp. 161-182.

Einstein on the Beach. By Philip Glass. Directed by Robert Wilson, 1976.

Hersch, Matthew H. "High Fashion: The Women's Undergarment Industry and the Foundations of American Spaceflight." *Fashion Theory,* vol. 13, no. 3, 2009, pp. 345–370.

Kiefaber, David. "Astronauts are the Stars in Louis Vuitton Ad," *Adweek, Adweek Blogs,* June 2, 2009. https://www.adweek.com/creativity/astronauts-are-stars-louis-vuitton-ad-14094/.

Lathers, Marie. "'No Official Requirement': Women, History, Time, and the U.S. Space Program." *Feminist Studies,* vol. 35, no. 1 (Spring 2009), pp. 14–40.

Lunden, Elizabeth Castaldo. "Barbarella's Wardrobe: Exploring Jacques Fonteray's Intergalactic Runway." *Film, Fashion, and Consumption,* vol. 5, no. 2 (2016), pp. 185–211.

Mears, Patricia, et al., *Expedition: Fashion From the Extreme.* Fashion Institute of Technology, Thames & Hudson, 2017.

Newman, Dava. Email to the author. March 31, 2019. See also Sara Beckmann, "Dava Newman presents 3D Knit BioSuit at 2022 MARS Conference," Post, MIT Media Lab, March 30, 2022.

Pollock, Steve. *Deadly Turbulence: The Air Safety Lessons of Braniff Flight 250 and Other Airlines, 1959–1966.* McFarland, 2014.

"Queen of Outer Space." *Variety* 31 Dec. 1957. Web. https://variety.com/1957/film/reviews/queen-of-outer-space-1200419077/.

Solomon, Matthew, ed., *Fantastic Voyages of the Cinematic Imagination: Georges Méliès's Trip to the Moon.* SUNY Press, 2011.

United States. House. *Hearing Before the Committee on Science and Astronomics.* 87th Cong., 2nd sess. 1962.

Weitekamp, Margaret A. *Right Stuff, Wrong Sex: America's First Women in Space Program.* Johns Hopkins UP, 2005.

Wosk, Julie. *My Fair Ladies: Female Robots, Androids, and Other Artificial Eves.* Rutgers UP, 2015.

Wosk, Julie. *Women and the Machine: Representations from the Spinning Wheel to the Electronic Age.* Johns Hopkins UP, 2001.

Women's Wear Daily Staff, "André Courrèges: Space Age Couturier." *Women's Wear Daily,* 10 Jan. 2016. Web. https://wwd.com/fashion-news/designer-luxury/andre-courreges-space-age-couturier-10307711/.

Zraick, Karen. "NASA Astronauts Complete the First All-Female Spacewalk," *New York Times,* 18 Oct. 2019, p. A11.

7

"His Own Personal Adventure"

Lunar Exploration and the IMAX Experience in *Magnificent Desolation* and *First Man*

ALLISON WHITNEY

Damien Chazelle's 2018 film, *First Man*, a biographical account of Neil Armstrong's journey to the Moon, presents its lunar exploration sequence in IMAX, using the large format's size and high definition to convey a remarkable and exclusive moment in human experience. The special properties of the IMAX sequence, and the extraterrestrial space it represents, are especially visible because up until this point, *First Man* was shot primarily on Super 16, a much smaller and grainier 16mm film format. The IMAX footage concludes with the spacecraft launching from the lunar surface, accompanied by a voice-over of Walter Cronkite and Eric Sevareid's television commentary. Sevareid begins, "I suppose they can never really describe to us, or will we ever know?" Cronkite replies, "No, and it may not be a beauty that we can pass on to future beholders either. These first men on the Moon have seen something that men who follow will have missed." Sevareid concludes, "They've peered into another life that we can't follow." In the final IMAX moments, the Moon leaves the frame and the image finally goes dark as another journalist's voice quite literally brings the story back to Earth with, " . . . Bob Simon at Trafalgar Square, where he joined thousands who watched an immense television screen as the Apollo touched down on the Moon" [*sic*]. The next shot is a low-definition television transmission of a crowd, themselves watching television, with an announcer stating, " . . . the average Romanian thinks of Apollo 11 a little bit as his own personal adventure." What follows is a brief montage of television coverage, focusing on the responses of broadcast audiences all over the world, before cutting to Armstrong (Ryan Gosling) and Buzz Aldrin (Corey Stoll) back on Earth in their quarantine facility, surrounded by newspaper and magazine reports of their journey. In selecting these archival television and audio clips to transition from the high-definition IMAX sequence back to the grainier terrestrial realm, Chazelle positions the distinct properties of audio-visual media (grain, noise, framing, static, timbre) at the center of a larger challenge: what can be represented, both of and through, the subjective sensory, intellectual, and emotional experiences of the Apollo astronauts? And how might one balance that experience with the heavy symbolic and historical complexities of that period in space exploration?

Every representation of the Apollo program has grappled, to some extent, with such challenges. Mark Cowen's 2005 IMAX documentary, *Magnificent Desolation: Walking on the Moon 3D*, opens with a collage of film and television images representing moments in both American and Soviet space history. While historical footage is common in space documentaries, this sequence utilizes IMAX 3D in a distinct way, such that each image seems to float into the frame from above, behind, or below the viewer, coalescing into a cloud that eventually forms a composite image of the Moon. This technique not only introduces the viewer to the sensations of 3D immersion, particularly as they operate on such a large screen, but also uses that visual language to demonstrate the complexity of the narrative, where thousands of interdependent achievements would culminate in the Apollo missions. At the same time, it suggests that the idea of "the Moon" is as much a manifestation of the public imagination as a celestial body, and the film goes on to acknowledge the roles of mythology, science fiction, folklore, and the history of astronomy in how we conceptualize the Moon and the implications of our visiting it.

The film then proceeds to the main attraction, a series of reenactments and dramatizations where astronauts explore a simulated lunar landscape, itself generated from photographs taken on the Moon. These episodes are presented within the rhetorical framework of IMAX, rendering the experience for the viewer in ways that seem tangible and grounded in the physical act of walking. As Alison Griffiths explains in "Time Traveling IMAX Style: Tales from the Giant Screen," IMAX is closely associated with the concept of virtual travel, both because the travelogue sub-genre corresponds easily with educational filmmaking, and because it so often evokes the "phantom ride," a cinematic trope where the camera travels on a vehicle exploring a landscape (Griffiths 242). The film's subtitle, "Walking on the Moon," emphasizes the embodied experience of walking even more than the vehicular thrills, while its use of 3D both simulates and extends stereoscopic vision, while drawing attention to the physical foundation of the viewer's senses. The film then uses that sensory engagement to cultivate an explicit subjectivity, placing the viewer in the position of the astronaut exploring the landscape, even to the point, I will argue, of effacing individual astronauts as historical figures.

This essay will reflect on these two cinematic treatments of the Apollo missions produced in IMAX. First, it will detail how filmmakers draw upon the specific properties of IMAX, using its size, image resolution, 3D capabilities, and associations with narratives of exploration and virtual travel to generate a dialogue with the other media that have documented, transmitted, preserved, and shaped public consciousness of space history. Filmmakers utilize the specifically immersive properties of IMAX, drawing upon a larger discourse of bodily engagement, subjective experience, and physical and sensory immersion, and they deploy these ideas to support larger thematic, educational, and expressive goals.

To begin, what exactly is meant by IMAX? The IMAX brand name refers to a great number of technologies, including film stock, film and digital cameras, and multiple theatre designs and projection systems. IMAX film is sometimes called 15/70, indicating 70mm film with 15 perforations per frame (as opposed to five perforations for standard 70mm), creating an image with a 1.43:1 aspect ratio. This aspect ratio, where the screen is 1.43 times as wide as it is tall, is much closer to a square than the wider ratios typically used in mainstream cinema such as 1.85:1 or 2.35:1, and as a result, IMAX is arguably as much a "tallscreen" medium as a widescreen one.

While the dimensions of IMAX screens vary, traditional theatres, usually housed in institutional facilities like museums and science centers, have flat screens between six and eight stories in height, or in some cases, semi-spherical IMAX Dome screens, originally termed OMNIMAX. The projected image is roughly ten times the size of a conventional film, occupying all or most of a viewer's field of vision, including the upper and lower portions, which are especially significant for one's feeling of orientation in space. IMAX uses a six-channel (and in newer facilities, twelve-channel) uncompressed sound system that allows for very precise directional effects. While both *Magnificent Desolation* and the lunar sequence in *First Man* were shot on IMAX 15/70 film, the filmmakers also use footage shot in other formats. *Magnificent Desolation* tends to show archival film and television footage in smaller frames within the full-screen image, while the

Super 16, 35mm, and television portions of *First Man* were blown up to IMAX proportions using a digital remastering system.

While digital 3D systems are now ubiquitous in mainstream movie theatres, for over twenty years, IMAX was the only format to offer consistent 3D film production and exhibition. In many respects IMAX was an ideal medium for pre-digital 3D filmmaking, as the large frame allows for complex compositions of both emergence and depth, while the stability of the projection system minimizes image vibration and thereby increases viewer comfort. *Magnificent Desolation* was projected on film and exhibited in 3D, but also in 2D when shown in IMAX Dome theatres. Further, it was released at a time of transition in the industry when IMAX was expanding to the multiplex market, adding to the existing network of museums and science centers. Meanwhile, by the time of *First Man*'s release, most viewers would see the IMAX version in digital projection, either the 4K IMAX Digital or 8K IMAX Laser system. Regardless of the specific type of IMAX projection, *First Man*'s shift to 15/70 footage is announced by a change in aspect ratio, occupying the full height of the screen. *First Man* was also screened in conventional cinemas, and in those cases, the shift between Super 16 and IMAX footage would have entailed a visible change in image quality and cinematography, but not aspect ratio. For the purposes of this essay, the focus is on *Magnificent Desolation*'s 3D exhibition, and *First Man* as it appears in an IMAX facility with its proportionally taller screen—where the differences among film formats would be most explicit.

The Photographer-Explorer

In *Magnificent Desolation*, immediately after the cloud of archival images coalesces into a composite Moon, the film proceeds to its first reenactment of Apollo 11 astronauts walking on the lunar surface. While Armstrong's first step might seem the logical starting point for this story, the filmmakers choose to begin at a later moment in the mission timeline. The camera tilts down from the sky to the landscape, whereupon an actor portraying Armstrong steps into the frame, his boots taking up much of the image. The camera then moves up his body to reveal the Hasselblad camera mounted on his chest, followed by his first lines of dialogue, where he asks Houston if they are receiving a television signal. He then walks toward the LEM (Lunar Excursion Module) and positions himself to photograph Aldrin descending the ladder. His camera audibly clicks, thus drawing viewers' attention toward the mechanism, while the scene is intercut with shots from Aldrin's point of view, using a fish-eye lens to denote the shape of his helmet, and archival photographs of Aldrin—the very ones that Armstrong is taking. As Aldrin steps onto the Moon, he almost immediately comments on the view, and delivers his observation of its "magnificent desolation." The Apollo 11 story is full of opportunities for beautiful, exciting, and inspiring imagery, so it is conspicuous that *Magnificent Desolation* begins with astronauts actively engaged with and commenting on the photographic and televisual media that document their journey, and visual elements of their own experience. Indeed, the film continually emphasizes such moments, presenting the act of not only looking but taking pictures as central to the exploration narrative; anchoring the film's claims of authenticity in historical photographs.

In order to appreciate the implications of photography's place in *Magnificent Desolation*'s narrative, it is important to consider the larger context of early twenty-first-century visual culture. In 2003, IMAX debuted a process called DMR, where conventional 35mm films are digitally remastered and blown up to IMAX proportions. This system was part of a corporate strategy to expand the theatre network while also encouraging dramatic filmmaking in the IMAX format. The first film to undergo DMR was Ron Howard's *Apollo 13* (1995), a film whose subject matter was well-suited to the educational mandates of the museums that housed the majority of theatres. Since then, it has become a common practice for mainstream productions—usually spectacle-heavy genre films—to have an IMAX release, and these versions are consistently promoted with the promise of an enhanced, or more authentic means of experiencing the films, and by extension, the spaces represented in those films. A current promotional slogan for IMAX is, "Watch a movie . . . or be part of one," the implication being that IMAX viewing entails a more active mode of spectatorship consistent with feelings of immersion. The advertising campaign for *First Man* included posters with slogans like, "Take the First Step in IMAX," or "Experience *First Man* to the Fullest," and specifying "Lunar sequences shot with IMAX cameras," indicating that the best way to see the film was in an IMAX facility, where the 15/70 footage could have its intended effect.

At the same time, the early twenty-first-century proliferation of digital filmmaking techniques and technologies, most evident to audiences in the form of CGI visual effects, contributed to an ongoing discourse on the relative realism of digital and analog image systems. In short, visual culture grappled with a tension between the photorealism of CGI and the material "authenticity" of celluloid film, based on the perceived indexical advantage of photochemical systems. To be clear, the notion that there are simple distinctions among these technologies or that any of them has a monopoly on realism is both specious and oversimplified, but the perceived tension between digital and analog media remains a powerful concept—one that IMAX exploits in its promotional rhetoric and in the visual logic of IMAX films. Even though the IMAX Corporation has begun to use digital camera and projection systems, IMAX 70mm film is still heavily coded as not only an alternative to, but a form of resistance against digital filmmaking, and directors like Christopher Nolan have made a point of framing their use of IMAX in these rhetorical terms (Whitney 32). While the thirteen years between *Magnificent Desolation* and *First Man* saw major shifts in visual technologies and the discourses around them, both films employ IMAX's rhetoric of realism in their claims of authenticity and underline the importance of their medium by dramatizing astronauts' acts of photography.

In *Magnificent Desolation*, the lunar landscapes are computer-generated, allowing for a speed and versatility of apparent camera movement that visualizes the Moon in a way that was not even available to the astronauts who visited it. Further, the filmmakers use CGI techniques in the reenactment scenes to represent phenomena that would be nearly impossible to photograph by conventional means, such as the movement of dust in an airless environment with 1/6 gravity, a crucial element when one is trying to make a realistic portrayal of astronauts' boots interacting with lunar material (Goldman 14). At the same time, the film's production suggested a unique affinity between original photographs taken by astronauts and IMAX film, as the negatives from the Hasselblad cameras used on the Moon were almost the same size and resolution as 15/70 stock and thus provided a level of visual detail that could approach IMAX standards (Chavanne 53). By mapping the astronauts' media apparatus onto the particular powers of IMAX cinema, the film makes a claim for a privileged experience of the Moon by at once using the latest digital special-effects technologies, including 3D animation, while anchoring the spectacle in the perceived authenticity of historical celluloid film.

While both the promotional discourse around IMAX and the emphasis on photography call attention to the process of mediation, it is also important to consider how the specialized exhibition space of IMAX cinemas offers its own rhetorical framing. For example, one of IMAX's particular powers is that it uses the large screen to enact dramas of scale. Theatres are typically designed so that audiences enter at the foot of the screen, where it towers over the viewer and sets the stage for subject matter involving large objects and vast spaces. This theatre design also encourages viewers to contemplate their own bodies and those of fellow audience members as scale markers, both for the screen and the objects projected as images. Charles Acland has argued that this configuration of the theatre, in addition to the common practice of briefing the audience on IMAX technology before screenings, means that, "unlike conventional cinema, it is impossible to forget you are watching IMAX technology" (431). That consciousness of the apparatus can then extend to expressive shifts in aspect ratio and image position in IMAX films, as the screen is large enough to offer complex multi-image compositions. In *Magnificent Desolation*, the filmmakers use each of these components of the IMAX experience to offer a narrative of exploration and adventure where the astronauts and the viewers alike gain a privileged vantage point, but also struggle to perceive and interpret the scale of the space around them.

One of the film's most astonishing sequences takes place at Hadley Rille, a huge lunar valley, rendered through digital animation and presented with a fast-moving aerial perspective. The sequence begins with a quotation from Apollo 17's Jack Schmidt, who describes experiences of confusion and disorientation in the alien landscape, where the lack of atmospheric cues or reference points like trees made it very difficult to judge distance and scale. As we listen to Schmidt's explanation, the film offers images of uninhabited landscapes, each announced by the click of a camera, followed by an astronaut figure dissolving into view to provide a marker of scale. In each case, the appearance of a figure in the landscape makes seemingly unremarkable spaces suddenly astonishing, revealing surprisingly large boulders, hills, and craters. This sequence then turns to a reenacted scene from the Apollo 15 mission, where Dave Scott and Jim Irwin are collecting samples at Hadley Rille. A voice from mission control inquires if they have found the lip of the rille, and the astronauts reply with hesitation at their inability to make out the topography. In this moment, the audience shares their confusion, as the camera's point of view seems to show an undifferentiated landscape. The voice from Houston continues, explaining that from the point of view of the television camera on the rover, it looks like they are standing at "the edge of a precipice." The television camera then emits a whirring sound as it turns, drawing audience attention toward it. As the astronauts finally realize their proximity to the edge, the camera tilts over the cliff, accompanied by dramatic music. This kind of shot is typical of IMAX documentaries and elicits a powerful sensation of vertigo. From there, the virtual camera swoops away from the astronauts to reveal the scale of the surrounding landscape, while the voice-over asks the viewer to imagine "standing at the precipice of a hole as deep as the Grand Canyon and not knowing it." The sequence concludes with an animation of the Statue of Liberty dwarfed by the walls of the rille. Again, this sequence utilizes IMAX's potential to represent scale, in concert with 3D, in order to demonstrate how the Moon challenged astronauts' perceptions, while at the same time anchoring the exploration narrative in the specific powers of visual technologies. The people on Earth watching the television feed are able to perceive the risk better than the astronauts themselves, while the combination of digital animation and IMAX 3D exhibition allows the mysterious Hadley Rille to become intelligible to the viewer in a way that eluded the astronauts. While *Magnificent Desolation* certainly encourages the viewer to identify with the figure of the astronaut, it also aligns the viewer with photographic technologies, both those that traveled to the Moon with the astronauts and the larger IMAX apparatus that makes the viewer feel they are "part of the movie."

First Man also uses the connection to analog photography and an emphasis on on-screen cameras to anchor the text in historical reality, while also suggesting that the IMAX apparatus offers a mode of sensory and emotional experience that parallels the astronauts'. The bulk of *First Man* was shot on Super 16, creating a markedly grainy image, with heavy use of shaky, handheld cameras. In the era of high-definition digital media, these aesthetic choices ground *First Man*'s appeals to historical and emotional authenticity in its image quality. These cinematographic properties allow *First Man* to resemble historical footage, as similar equipment was used to document many NASA activities during the Gemini and Apollo programs, to say nothing of the proliferation of 16mm films distributed for educational and promotional purposes in that era. The small size of Super 16 equipment offered versatility in camera movement and placement inside the cramped spaces of air- and spacecraft. While movie sets are designed with camera placement in mind, the smaller 16mm cameras allowed the cinematography to achieve tight angles and extreme close-ups that not only generate feelings of claustrophobia but also encourage close attention to the minutiae of both the intricate machines and the actors' performances.

These techniques evoke a particular brand of realism that is quite distinct from the glossy CGI one might expect from a space exploration movie released in 2018.

First Man establishes an aesthetic language where the cinematography prioritizes the emotional, internal, and psychological perspectives of the characters using a visual baseline of grainy close-ups, sometimes verging on abstraction. The IMAX sequence begins at the moment the astronauts open the door of the LEM, the size of the frame expanding vertically (from a 2.39:1 aspect ratio to 1.43:1) just as the camera swoops out the door and then stops to gaze upon the landscape. The now-IMAX point of view pauses, giving the viewer time to adjust to a new world and a new way of seeing, characterized by hyperclarity, vertical immensity, and the smooth movements of crane shots and the Steadicam. In their commentary track on the Blu-ray release of *First Man*, Chazelle and cinematographer Linus Sandgren explain that they made a point of using Steadicam shots in the lunar sequence exclusively, making it even more formally distinct from the rest of the film. In a promotional video entitled "*First Man* IMAX Behind the Frame," Chazelle describes this as a "*Wizard of Oz* moment," comparing his shift in format to the transition from sepia to Technicolor as Dorothy crosses the threshold in the 1939 film. This is a fitting comparison, as it is a moment where the emotional impact on the characters is announced through a deployment of specific film technologies and in a manner that the audience is meant to notice.

Even though *First Man* goes to enormous lengths to portray the events from Armstrong's subjective position, the IMAX camera acquires a level of autonomy, even departing the LEM well before the astronauts. The second shot in the sequence is from a low angle, looking up at Armstrong and Aldrin still inside, pondering their first steps into a new world whose aesthetic parameters are aligned with the technical specificities of IMAX. Once again, as in *Magnificent Desolation*, the properties of IMAX are used to denote sensory and cognitive elements of the astronauts' experience, while also operating on a more metaphorical level, where the "new ways of seeing" afforded by the film technology correspond to the much broader historical and conceptual transformations that hinged on the Apollo missions.

The remainder of the Moon sequence is shot in IMAX, and even in the brief cuts to Super 16 flashbacks and television footage, the IMAX aspect ratio persists, literally re-framing both Armstrong's memory and the collective experience of Apollo 11 as a mass media event. As Armstrong prepares to climb down the ladder, the film details the deployment of the television camera that will record his first steps, with Armstrong's movements reflected in its lens. While *First Man*'s emphasis on cameras is not as persistent as in *Magnificent Desolation*, the film does make a point of aligning the television apparatus and the IMAX camera that photographs it, and it does so most dramatically through its sound design. When the IMAX camera moves out the door, there is a whooshing sound of air rushing out of the LEM, followed by silence. The silence persists through the deployment of the TV camera, and sound only returns to the sequence once we cut to Armstrong's point of view and hear him breathing inside his helmet.

Complete silence is very rare on film soundtracks, but its placement here is much more than a matter of realism in a world without atmosphere. First, the silence gives the audience an opportunity to focus on the change in image quality, and to contemplate the otherness of the landscape in a mode of awe and wonder. Second, the emphasis on the television camera, accompanied by a brief insert of its broadcast of Armstrong, reminds the audience of the viewers on Earth and their own sense of astonishment, while at the same time offering a stark contrast with IMAX image quality. The ensuing dialogue with mission control includes their remarking that they can see Armstrong on TV, as well as instructions for Aldrin about setting up camera equipment. As Armstrong assesses the ground below the lander, he describes it as "very, very, fine grain . . . almost like a powder." While this might seem a mundane observation, and a curious one to include in this dramatization, the comment draws attention to the fact that the fine grain is indeed visible in IMAX, again demonstrating the medium's affinity for the particular qualities of this alien landscape. After his first step and "giant leap" statement, the silence briefly returns, with point-of-view shots of Armstrong looking down at his walking feet, followed by a 360-degree panoramic shot of the landscape. In the midst of this panorama, the theme music begins, and it will persist for the rest of the sequence. Much in the way that sound emphasized photography in *Magnificent Desolation* by drawing attention to cameras, silence seems to have a comparable function in *First Man* by reinforcing the tactility and visuality of the experience and positioning the aesthetic characteristics of IMAX as uniquely suited to convey the subject matter. While it is true that these details have a historical foundation, as there was indeed a television camera and Armstrong did comment on the texture of the ground, Chazelle's decision to include them in *First Man* capitalizes on the aesthetic distinctions offered by its film formats. In the visual and sonic language of the film, the Moon both requires and elicits new modes of perception and representation.

His Own Personal Adventure: IMAX and the Astronaut

Magnificent Desolation seeks to inspire its viewers to pursue space science by asking them to imagine themselves as the figure of the astronaut. Indeed, the film's final episode is an imagined future where a child who appeared in an earlier sequence, Veronica Lugo, envisions herself as an adult (Brandy Blackledge) living on a Moon base. In order to appreciate how the film sets up the viewer's relationship to the astronauts on screen, it is important to consider the larger history of IMAX space documentaries. As Valerie Neal explains in "Bringing Spaceflight Down to Earth: Astronauts and the IMAX Experience," there is a whole sub-genre of IMAX devoted to space exploration, including productions by IMAX Space Ltd., a division of the IMAX Corporation formed to foster a collaboration with NASA to design cameras for use in space, train astronauts as IMAX cinematographers, and create films that inspire space science.

Fittingly enough, it was Apollo 11's Michael Collins who, as inaugural director of the Smithsonian's Air and Space Museum, suggested that IMAX cameras be sent to space and that the Museum install an IMAX facility, arguing that this medium was well suited to representing the experience of space travel (Neal 152). Neal describes a gradual shift in the personification of astronauts in these documentaries, where the earlier films tended to present them nearly anonymously, as "typical shuttle astronaut[s]" rather than as individuals, while later productions highlighted astronauts' distinct personalities (162). She explains that the filmmakers effected this evolution primarily through their use of narration and dialogue. Earlier productions like Graeme Ferguson's *The Dream Is Alive* (1985) used voice-over to describe the astronauts' activities, but their own voices were rarely heard, while later films like Toni Myers's *Space Station 3D* (2002) were largely narrated by the astronauts themselves and included substantial commentary on their personal experiences (Neal 165). To be clear, *Magnificent Desolation* was neither an IMAX Space production nor was it shot in space, but it inherits many of those films' strategies for cultivating the figure of the astronaut, particularly through the use of voices. As I will explain, one of the consequences of the film's strategies is that even as it highlights the experiences of real people who visited the Moon, it also has a curious tendency to de-historicize them, to at once use their individual experiences and testimony while dispersing their individual identity to create a point of identification for the viewer.

Magnificent Desolation's narrative structure rarely differentiates one lunar mission from another, organizing the material around common astronaut experiences rather than tracing each mission as a discrete historical event. While the actors do reenact specific events, the film jumps back and forth in the timeline such that the viewer quickly loses track of which figure is which. The film actively superimposes and conflates different moments in lunar history, sometimes quite literally through the use of dissolves, where astronauts continually appear and disappear from an otherwise stable landscape. For example, a sequence detailing the evolution of lunar transportation explains how the Apollo 11 and 12 astronauts could only cover a few hundred yards on foot, while Apollo 14 introduced the MET (Modular Equipment Transport), allowing them to move equipment over larger distances, and then finally the Apollo 15, 16, and 17 astronauts could travel even farther afield by driving the lunar rover. These innovations are shown in sequence, with the respective astronauts appearing and vanishing from the landscape, each one progressively further from the camera to denote the distance traveled. Finally, astronauts from multiple missions appear in the landscape simultaneously, thus effacing historical distinctions entirely. Even in scenes focusing on a specific individual, such as Charlie Duke's account of leaving his family portrait on the Moon, the actor portraying him first photographs the picture, continuing the film's trope of photography, but as he turns to walk away he vanishes, leaving an empty landscape. This persistent visual effect makes an important point about temporal scale, contrasting the deep time of the landscape with the humans' brief visitation, but it also has a curious effect on the experience of embodiment—the dematerialization of the astronauts as specific historical figures creates a space for the viewer to imagine themselves in an abstracted astronaut position, one that is encouraged by the particular visual powers of IMAX 3D.

The film also supports its discourse of embodiment through strategic use of sound, particularly voices. As a public face of the space program, Apollo astronauts routinely offer accounts of their experiences, which have been recorded and shared through interviews, memoirs, and media appearances. Further, as Neal's research indicates, repeat viewers of IMAX space films would be familiar with astronaut narration in that particular medium. *Magnificent Desolation* makes extensive use of astronaut testimony, but apart from a few instances where we hear archival recordings, most of the accounts are performed in voice-over by well-known Hollywood actors, including Morgan Freeman as Neil Armstrong, Bryan Cranston as Buzz Aldrin, Matt Damon as Alan Shepard, Matthew McConaughey as Al Bean, John Travolta as Jim Irwin, Bill Paxton as Ed Mitchell, and Gary Sinise as Gene Cernan, among others. Further, the film's principal narrator is Tom Hanks, who not only portrayed Jim Lovell in Ron Howard's *Apollo 13* but also narrated the television series *From the Earth to the Moon* (1998). The decision to use not only actors, but Hollywood actors with recognizable voices, can, on the one hand, intensify the emotional and dramatic weight of the testimony, while on the other, distance that testimony from its original authors. While each quotation is immediately attributed to the astronaut who said it, there is a disconnect between the historical recordings, the reenactors' dialogue, and the more famous voices heard through voice-over. For example, Dave Scott from Apollo 15 is portrayed in three different forms: his own voice in historical recordings, the actor who portrays him on screen in the reenactments, and Paul Newman's voice-over recitation of Scott's recollections. The film thereby produces a dynamic where the figure of the astronaut on screen is both invested with the historical and emotional weight of the real-life astronauts' accumulated experiences, while also rendered abstract enough to allow viewers to imagine themselves as space explorers. This strategy is consistent with the film's educational and promotional agenda, as it is not so much a history lesson as it is an appeal to the senses, and one that argues for the pursuit of science as a search for personal experiences of wonder.

While *Magnificent Desolation* condenses multiple people and performances into an abstracted astronaut figure, *First Man*'s strategies of immersion and embodiment conform tightly to Armstrong and his subjectivity. Indeed, the very first shot of *First Man* shows Armstrong piloting the X-15, but lit in such a way that his face is invisible behind his darkened helmet. The rest of this sequence is dominated by shots from Armstrong's point of view, extreme close-ups of his face, and small portions of the aircraft's exterior, establishing a visual language markedly different from most dramatizations of space travel, and focused explicitly on Armstrong's physical, intellectual, and emotional experience. Even in the lead-up to the Apollo 11 launch, the film reconceptualizes typical rocket launch imagery by presenting it in explicitly subjective terms. For example, documentaries and dramatizations alike almost always use "rocket as seen from the elevator" shots, where the camera's upward movement emphasizes the enormity of the vehicle. *First Man* offers such a shot, but the lighting and camera position make a unique intervention, ensuring that the astronauts' reflections are highly visible in the elevator's window, rendering even the scale of the Saturn rocket an opportunity for personal contemplation. *First Man*'s preoccupation with Armstrong's sensory, psychological, and cognitive relationship to this mission is also communicated through the use of flashbacks, drawing clear connections between each part of the mission and Armstrong's personal history. These flashbacks are specifically overlaid with grief, and in moments when Armstrong experiences physical and psychological duress, such as difficult training procedures or suspenseful moments in space missions, the film will cut to flashbacks of his daughter, Karen, whose death from cancer is detailed early in the film.

On the surface, the panoramic properties of IMAX, as well as the smoothness of Steadicam movement, as contrasted with the visceral associations of the hand-held camera and constant close-ups in the Super 16 footage, might seem to estrange the IMAX sequence from the subjectivity and intimacy of *First Man*'s visual language. And yet, the film manages to utilize IMAX to present the experience of walking on the Moon as yet another transformation of the subject. While the camera does seem to have a level of autonomy, sometimes moving independently of the astronauts, it also immerses us in the alien space that will allow Armstrong a transcendent experience. In an earlier sequence where Armstrong is being interviewed as a candidate for the Apollo program, he answers the question of why space exploration is important by saying, "When you get a different vantage point, it changes your perspective . . . it allows us to see things that maybe we should have a long time ago, but just haven't been able to until now." In this case, the shift to IMAX offers a transformed way of seeing, not only of the lunar landscape but also of the psychological interior. Indeed, in that same interview, the NASA officials also ask Armstrong if the death of his daughter has affected him—a strange question, but one that the lunar sequence ultimately addresses. As Armstrong walks on the Moon, he again has flashbacks to his family and Karen, but in this case, the Super 16 footage is blown up to IMAX proportions. These flashbacks mark the dramatic aesthetic differences between IMAX and Super 16, the latter looking especially grainy and with deeply saturated colors in contrast with the clarity and gray scale of the lunar palette. The fact that they appear in the same aspect ratio demonstrates that Armstrong's memories are literally reframed by the Moon, viewed with a changed perspective, by an experience of almost incalculable intensity that, in Sevareid's words, "they can never really describe to us." Armstrong experiences a series of these flashbacks, many of them showing his family in a lush green landscape, intercut with the stark lunar one, before he pauses at the edge of a crater. In a fictionalized moment, he pulls out a bracelet that belonged to Karen, dropping it into the crater where it falls out of focus and disappears into the darkness. The sequence ends with his turning away from the crater and walking off-screen, followed immediately by the final IMAX shot, as he and Aldrin begin their journey back to Earth.

* * *

While *Magnificent Desolation* and *First Man* have very different tonal properties and modes of filmmaking—from educational documentary to impressionistic character study—their uses of IMAX, in its scale, image quality, and rhetoric of immersion, represent the Moon in highly subjective terms. At the end of *Magnificent Desolation*, in Veronica's imagined future as an astronaut, we see her walking in her spacesuit and pausing to reenact the moment when Charlie Duke photographs his family portrait, but here the picture she places on the ground is her childhood painting of a spaceship. The painting then comes to life through animation, the spaceship transforming into a rocket heading toward Mars. The sequence then concludes in a manner reminiscent of the film's opening, with a cloud of images of subsequent space explorers floating around the viewer, this time coalescing into a composite image of an astronaut, followed by a quotation from Socrates: "Man must rise above the Earth to the top of the atmosphere and beyond, for only then will he fully understand the world in which he lives." *First Man* uses IMAX, in dialogue with 16mm, to internalize that goal of understanding, using the intense subjectivity of immersive cinema to align the aesthetic novelty of the lunar landscape, and the experience of walking upon it, with the existential transformation of not only a man, but of mankind. *Magnificent Desolation* and *First Man* actively demonstrate a connection between the desire for new perspectives, from the aerial and distant one achieved through space travel to the intimate and tangible detail afforded by film technologies, and the chance for reflexive introspection. Both films promise their audiences a chance to identify with the exclusive experience of the handful of people who have visited the Moon, and to do so first by aligning IMAX technologies with cameras and photographic views within the films, and then by asserting that IMAX's discourse of enhanced vision, sensory immersion, and virtual travel can render iconic historical and scientific events in the terms of a "personal adventure."

Works Cited

Acland, Charles R. "IMAX Technology and the Tourist Gaze." *Cultural Studies*, vol. 12, no. 3, 1998, pp. 429–445.

Apollo 13. Directed by Ron Howard, Universal Studios, 1995.

Chavanne, Bettina. "Ticket to Moon Walk." *Smithsonian Air & Space*, vol. 20, no. 4, 2005, pp. 52–57.

The Dream Is Alive. Directed by Graeme Ferguson, IMAX Space Ltd., 1985.

First Man. Directed by Damien Chazelle, Universal Pictures, 2018.

"*First Man* IMAX: Behind the Frame" *YouTube*, uploaded by IMAX Corporation, 10 October 2018, www.youtube.com/watch?time_continue=89&v=hJCg_tq16Wg.

From the Earth to the Moon. Produced by Tom Hanks, Brian Grazer, Ron Howard, and Michael Bostick, HBO Original Programming, 1998.

Goldman, Michael. "Clips: Film: Moon Dust." *Millimeter—The Magazine of Motion Picture and Television Production*, vol. 34, no. 5, June 2006, p. 14.

Griffiths, Alison. "Time Traveling IMAX Style: Tales from the Giant Screen." *Virtual Voyages: Cinema and Travel*, edited by Jeffrey Ruoff, Duke UP, 2006, pp. 238–258.

Magnificent Desolation: Walking on the Moon 3D. Directed by Mark Cowen, Playtone, 2005.

Neal, Valerie. "Bringing Spaceflight Down to Earth: Astronauts and the IMAX Experience." *Spacefarers: Images of Astronauts and Cosmonauts in the Heroic Era of Spaceflight*, edited by Michael Neufeld, Smithsonian Institution Scholarly Press, 2013, pp. 149–174.

Space Station 3D. Directed by Toni Myers, IMAX Space Ltd., 2002.

Whitney, Allison. "Cinephilia Writ Large: IMAX in Christopher Nolan's The Dark Knight and The Dark Knight Rises." *The Cinema of Christopher Nolan: Imagining the Impossible*, edited by Jacqueline Furby and Stuart Joy, Wallflower, 2015, pp. 31–43.

The Wizard of Oz. Directed by Victor Fleming, MGM Studios, 1939.

8

America's "Space Frontier" in an Era of Space Tourism

JAMES A. SPILLER

In February 2017, the commercial spaceflight company SpaceX announced plans to send fee-paying passengers on a lunar fly-by before the end of 2018. While industry observers (who were otherwise bullish about space tourism) questioned this timeline, they overlooked the symbolism of the expedition's target date (Chang 2017). Taking place during the fiftieth anniversary of the first lunar orbit by Apollo 8 astronauts, the SpaceX mission would have signaled the dawning of a new space age. This new phase of spaceflight would be very different from the one highlighted by *Time* magazine when it hailed the Apollo 8 astronauts as "[M]oon pioneers" exploring the so-called space frontier so that waves of American men and later women could follow ("Men of the Year"). This widely heralded national project was coordinated by the National Aeronautics and Space Administration (NASA) and defined by the productive labor of citizen space travelers committed to improving the lot of humankind. Although proliferating satellites have positively impacted the world since Apollo 8, widespread human spaceflight spearheaded by NASA has not unfolded. SpaceX and other commercial spaceflight companies now promote private enterprise as the pioneering vanguard and premium-cost tourism as a means to jump-start human travel to the Moon, Mars, and beyond.

Tourism did not play this essential role in earlier, Cold War–era roadmaps for human activity in outer space. When the Soviet Union surprised the world in October 1957 by launching the first satellite (Sputnik), Washington established NASA to achieve civilian leadership in space and boost US prestige by demonstrating its unique ability to develop space for "the benefit of all mankind" (Spiller 50). Conquering this forbidding realm, what Americans commonly called the "space frontier," was a lofty cause seemingly in line with the nation's pioneering past. During this first phase of human spaceflight that culminated in lunar landings, Americans often regarded astronauts as national envoys, courageous scouts paving the way for others to follow and conquer the space frontier. Although this frontier motif went dormant in the 1970s, it revived for a time during the era of the space shuttles, the second phase of US-piloted spaceflight. Shuttle astronauts came off in the 1980s as different types of pathfinders, educated professionals who were opening the space frontier to research and development (R&D), industry, and settlement. In both phases, these pioneering astronauts represented the nation as they advanced an epochal project that would see many more Americans living and working in space, but not as well-heeled tourists seeking mere personal pleasure.

Contemporary proponents of space tourism, who suggest that this industry can leapfrog a moribund NASA and launch a third phase of spaceflight, often use the motif of the space frontier to stimulate public interest in ways that may help to attract clients, private investors, and state and federal subsidies. When those proponents do so—including billionaire entrepreneurs who assert that space tourism will help kick-start that third phase so that people can improve life on Earth (Richard Branson of Virgin Galactic), explore and settle Mars (Elon Musk of SpaceX), or build space colonies (Jeff Bezos of Blue Origin)—they count on the cultural currency of a decades-old motif to inspire public support. Americans may be unmoved by such visionary boosterism, however, for the nationalist narrative of America's space frontier has lost much of its saliency. Furthermore, it is unlikely that this narrative, which traditionally celebrates egalitarian opportunity and collective effort on behalf of the nation and humanity, can be revived by a nascent space tourist industry organized to maximize shareholder profit and serve extremely wealthy clients who can afford expensive ventures into space.

Astronauts as Frontier Scouts

As it prepared to land men on the moon, NASA's scholastic booklet, *Space: The New Frontier*, explained that "the characteristic American confidence in the future" that "brought the first colonists westward across the Atlantic" and subsequent "generations westward across the continent" had prompted the nation to set off for outer space (NASA, *Space*). Students were likely receptive, for popular entertainment brimmed with tales of frontier derring-do in the Wild West and in space. Such media populated the cosmos with aliens whose taste for killing white men and abducting their women smacked of boilerplate "Indians" (Horrigan). These media sometimes dispensed with such frontier pulp and depicted in technical realism a process of cosmic exploration paralleling that of the New World and the American West. This tradition matured in the 1950s, as visionaries like Wernher von Braun, a German wartime missile engineer turned US Army rocket designer, depicted in technical detail an impending age of space travel. Just as caravels and covered wagons opened New World frontiers, von Braun predicted that pioneering rocketmen would prepare the way for waves of space travelers (McCurdy). Hollywood, popular periodicals, and Walt Disney (by way of television and amusement park attractions) introduced millions to this scenario, what Disney himself called the conquest of America's "new frontier, the frontier of interplanetary space" ("Man in Space").

When the US announced plans to launch the first satellite in the late 1950s, years of popular entertainment had thus prepared Americans to think of space as a new frontier, even if they did not anticipate pioneering it anytime soon (Michael). They certainly did not expect the Soviet Union to blaze a trail into space, but the USSR did just that in October 1957 by sending Sputnik into orbit. Many believed that frontiers were the wellspring of American dynamism, of its exceptional freedom, democracy, and prosperity, which in turn made it uniquely equipped to pioneer new frontiers in this age of advanced science and technology. As presidential adviser Vannevar Bush explained in his 1945 report, *Science: The Endless Frontier*, societal progress stems from technological innovation that "depends upon a flow of new scientific knowledge." He felt that this flow was strongest in America, due to federal investment in R&D and to the nation's frontier legacy of liberalism. This legacy gave the US an unassailable lead in science and made it inconceivable that the illiberal Soviet Union could catch up. As Bush confidently noted, "a totalitarian state cannot compete with a free people in the advancement of science, for the dictation and dogma are contrary to the free spirit of inquiry, which is the heart's blood of scientific advance" (Bush, *Science*).

Such confidence failed after Sputnik. Thus as the *New York Times* worried that people might mistakenly believe "that Moscow has taken over world leadership in science" ("Soviet Claiming Lead in Science" 2), US officials pursued a leading space program that would "be construed by other nations as dramatically symbolizing national capabilities and effectiveness" (Merchant 3). That program would generate "a public image of supremacy," a Congressman declared, and impress "peoples of the world of that reality of [American] power" (Van Dyke 396). A prominent science adviser agreed and counseled that, "In our day, when few physical frontiers remain, people visualize space" as a challenge "that must be accepted by a great nation to demonstrate its mettle" (Berkner). Federal officials heeded that counsel and created NASA in 1958 to demonstrate America's mettle with a preeminent civilian space program. NASA spokespeople and media publicists often echoed this discourse and described the US space program as a pioneering endeavor. They did so because the two strands of the frontier motif, those associated with historian-turned-president Theodore Roosevelt and his academic peer Frederick Jackson Turner, synergistically cast this costly program as true to the nation's historic character and its lofty aim of benevolent global leadership.

A half-century before the space age, Turner suggested that America was a uniquely free, democratic, and prosperous country because of its frontiers. These purportedly vacant lands called on the rugged individualism and democratic cooperation of pioneer settlers, protecting these virtues from the illiberal influences of the emerging industrial order. Further, since those successive frontiers offered productive outlets for surplus capital and labor mounting in American cities, they provided an essential stimulus for the US economy (Turner). Turner's frontier story of liberal vitality and national prosperity later resonated with the general optimism of post–World War II America. It also allayed Americans' anxiety after Sputnik and the April 1961 orbit of the first man—a *Soviet* man—which the USSR called the beginning of a "new era in human progress" made possible by "the great might of socialism" (*Soviet Man in Space* 12). The US space program was designed to answer such provocations and "demonstrate once again," in the words of NASA's first administrator, "that free men—when challenged—can rise to the heights and overcome the lead of those who build on the basis of subjugation" (Glennan). The Turnerian vein of the space frontier motif helped make that point since Turner's legions of pioneers demonstrated liberalism's vitality, fueled national prosperity, and laid the foundations for America's scientific and technological supremacy (Turner 31–32). NASA's proponents believed that the hundreds of thousands of people working to put astronauts into space would do the same and beat back Moscow's spacefaring challenge. Pioneering the celestial frontier demanded supreme performance that only a free and prosperous country could sustain, a national capacity then exemplified by the astronauts selected for NASA's first manned spaceflight initiative, Project Mercury.

The seven Mercury astronauts did not resemble Turner's bands of homesteaders. They better matched the trappers, scouts, and explorers exalted by Theodore Roosevelt as models for his enervated peers. Worried that industrial-age Americans had forsaken their pioneering paternity and racial mettle, Roosevelt urged them to pursue the "strenuous life" of muscular vigor and lofty purpose of those frontiersmen. Roosevelt warned that if Americans shrank from this noble effort, "then the bolder and stronger peoples will pass us by, and will win for themselves the domination of the world" (Roosevelt). Decades later, many Americans worried that materialism and conformity once again threatened their global stature and allowed the Soviet Union to surpass them in space. A prominent Catholic theologian warned of communist "barbarians, hardy and disciplined, ready to defeat a civilization in which the very achievements of its technology have made for soft and indulgent living" (Niebuhr 30). The Protestant evangelist Billy Graham similarly worried that Americans "are growing soft" and their "amusements and greed for money is acting as a sedative" [*sic*]. He channeled Roosevelt by urging Americans to "toughen up!" through "compulsory scientific training in all our schools, as well as compulsory physical training for all our young people" (Graham).

Graham may have found cause for optimism in the Mercury astronauts. As military officers who eschewed lucrative civilian careers, they became national heroes not only by choosing modest earnings and national service but also by exhibiting outstanding toughness and bravery. As novelist Tom Wolfe later wrote, "reporters and broadcasters dealt with [them] in tones of awe . . . that one has of an impending death-defying stunt" (Wolfe 71). The terrifying danger of riding rockets into space did not seem to shake these men, for they had already faced mortal peril as test pilots or combat aviators. Like Roosevelt's backcountry heroes, these courageous men—in the words of *National Geographic*—were also "superb physical specimens" who rose to the rank of Mercury astronaut after a "harrowing ordeal of physical and mental examination." The magazine documented that ordeal and declared each astronaut prepared for the daunting trials of spaceflight; "physically and mentally, this is an extraordinary man, and for years he has worn discipline like a shield" (Fisher 52, 59). As *LIFE* magazine reported, this discipline was critical because spaceflight "would expose them to greater stresses than most pilots had ever encountered," and they needed "nerves of steel" and an "unfailing instinct for making calm, steely, split-second decisions." Their manly heroism stemmed from this ability to "be the masters of their own destiny" (Carpenter et al. 6–7).

The historic May 1961 suborbital flight of Alan Shepard, in which this first spacefaring American took control of his capsule, established the astronauts' bona fides as masters of their own destinies. After John Glenn did the same months later, with his *Friendship 7* capsule during America's first orbital spaceflight, he and other Mercury astronauts were depicted as self-directed "helmsmen" who combined "the pioneering image of '150 years ago' with a forward-looking mastery of technological change" (Smith 200). A university president drew this very "analogy between the problems of the pioneers of the Western Frontier and the pioneers of outer space," affirming that "both ventures require personal fortitude of a high order, integrity, courage and perseverance." In his mind, the flight of *Friendship 7* proved that "the elaborate equipment of a Mercury Capsule is nothing without a John Glenn" (Rudder 19).

Although *Time* later admitted that "astronauts often seem to be interchangeable parts of a vast mechanism," the magazine insisted that they were in fact "essentially loners" prepared to be self-reliant in space ("On Courage in the Lunar Age" 19). Their heroic stature hinged on the gendered nostalgia that such men still existed and that they modeled "the manly traits," as one star-struck nun explained, that boys should "portray in their adult lives" (Glenn Jr. 3, 92). This stature also reflected chauvinist norms that treated men as plucky pathfinders and women as lacking the nerve and strength to pioneer the space frontier. De facto prohibition against female astronauts reinforced this prejudice, giving women little opportunity to prove that they could gracefully handle spaceflight. When the clinic screening the Mercury astronauts determined in 1959 that women could in fact handle the rigors of space travel, newspapers like the *Philadelphia Inquirer* largely dismissed the idea, joking that a "young unmarried woman . . . may make a fine test pilot for a simulated space ship. But wait until the simulation is gone, and with it all chance of seeing the boyfriend for an indefinite time, and watch the old pioneer spirit evaporate into space" (Weitekamp 91–117).

Like women, minorities found little in common with the Mercury astronauts and their successors who flew to the Moon. The astronauts embodied the civic ideal that their egalitarian nation would deliver new frontiers of freedom and prosperity to all humankind, but they were also Rooseveltian heroes—all white and, in the case of the Mercury astronauts, all Protestant. Theodore Roosevelt progressively welcomed diverse peoples into his idealized American melting pot, but he believed that Anglo-Saxon and Teutonic men had built that exceptional country. This strain of racial nationalism continued into the early space age, even as civil rights became pressing national concerns. The African American magazine *Ebony* proudly featured Black women who worked for NASA as number-crunchers and Black men who supported the astronaut corps as medical assistants, but it was not yet able to feature astronauts of color ("Space Doctor for the Astronauts" 35–39; "Angel of Mercy to the Astronauts" 49–52). Astronauts may have stood for America's gathering civic idealism, but they offered a comforting reference to Rooseveltian chauvinists even as the US became a more pluralistic society.

When this first phase of human spaceflight closed after the lunar landing program in the early 1970s, public discourse about astronauts and the space frontier changed as an increasingly pluralistic society rejected Rooseveltian chauvinism and as the cash-strapped United States dramatically downsized NASA. When the US spaceflight program entered its second phase in 1981 with the first flight of the space shuttle, astronauts evinced little of the manly heroism of the Mercury era. Owing to dramatic changes in American culture, political economy, and space technology, they came off instead as pioneer settlers with graduate degrees, poised to stimulate national competitiveness by developing the space frontier.

Shuttle Crews as Frontier Settlers

Time magazine took up the familiar trope of the space frontier when it designated its 1968 Men of the Year—the crew of the lunar-orbiting Apollo 8—"America's [M]oon pioneers." Like Theodore Roosevelt's frontier explorers, they displayed "courage, grace, and cool efficiency." By asserting that the "newer world opened up by the Men of the Year will surely, in time, reach far beyond the [M]oon," the magazine echoed Frederick Jackson Turner as well and imagined that subsequent waves of space pioneers would launch "a journey into man's future." Although it acknowledged that many Americans "question the wisdom of spending billions to escape the troubled planet," *Time* failed to forecast the waning fortunes of the space frontier motif, particularly the collapse of its chauvinistic pillars ("Men of the Year").

That impending collapse was evident in 1969, when an essayist opined about the astronaut corps: "We have demonstrated our belief that white-skinned, crew-cut types can do just about anything to which they set their minds. . . . Must we not demonstrate our belief that brown-skinned, black-skinned have similar capabilities?" (Brown). Several years later, an editorialist who regretted "that, in space, woman is still a joke," insisted that women had those capabilities as well (Nauton 17). After the press "raked NASA over the coals for its equal opportunity shortcomings" and federal officials pushed for minority and female recruitment, the space agency did just that and in 1978 selected six women and three Black men for the first cohort of space shuttle trainees (Scott). This new generation of diverse astronauts was depicted less as Rooseveltian conquerors than as Turnerian settlers, ready to railroad into a boundless frontier and thereby revive a flagging country.

That narrative emerged after a decade in which the dream of pioneering spaceflight faded, as Americans grew dispirited by their stricken economy, turbulent society, and endangered environment. On the very day the Apollo 11 astronauts returned from their historic lunar landing, NASA Associate Administrator George Mueller pitched that embattled dream. Just as Turner credited western frontiers for saving America from the ills besetting fully settled nations, Mueller extolled outer space as a safety valve from planetary confinement. He accordingly warned that if Americans were to forsake "the spirit of our forefathers then will man fall back from his destiny, the mighty surge of his achievement will be lost, and the confines of this planet will destroy him" ("Statement by George Mueller"). Vice President Spiro Agnew similarly urged Americans not to "turn inward, away from the opportunities and challenges of its most promising frontiers" as he promoted a post-Apollo program entailing a fleet of space planes, an orbital station, and Moon bases, and a piloted mission to Mars ("Statement by Vice President Spiro Agnew").

While NASA Administrator Thomas Paine lobbied hard for that program and Wernher von Braun insisted it would open the economy-boosting "new frontier of space," President Richard Nixon rejected this costly plan and determined that a retrenched NASA needed to focus on a "practical application" to solve the "many critical problems here on this planet" (von Braun 31–33; "Statement about the Future of the Space Program").

Throughout the 1970s, those earthly goals influenced the development of the space shuttles, which Nixon called "a wise national investment" that would help "reorient our national space program so that it will have even greater domestic benefits" ("Annual Message to the Congress on the State of the Union"). The *New York Times* recognized that critics dismissed the costly program as "another grave distortion of national priorities," but it endorsed the shuttles as "a major investment in the future" that would "alter the economics of space activities and provide dividends that should continue for decades to come" ("Investment in the Future").

NASA's new chief, James Fletcher, appeared to share this outlook when he avowed in 1972 that America was on track with a practical and cost-effective spaceflight program (Remarks by James Fletcher). Fletcher turned out to be an evangelist for the space frontier, however, and pushed back on NASA's bare-bones budget by declaring that "in concentrating on the 'now' problems we are forced to ask questions about the future: are we losing sight of 'the dream?'" (*NASA and the Now Syndrome)*. Whereas President Nixon had said the space shuttles would "help transform the space frontier" in the future after the US regained its financial footing (Statement by President Nixon), Fletcher took a page from Frederick Jackson Turner and reversed this timeline. He suggested that an immediate push into space would revive the nation's economy and spark a new industrial revolution. Fletcher's dream paralleled the imaginative proposals of physicist Gerard O'Neill. Recognizing public fear of national and environmental decline at that time, O'Neill insisted that America could overcome these post-frontier ailments by rapidly colonizing the "High Frontier." The solar utilities and space-based factories tended by his "booming frontier settlements" would deliver inexhaustible resources and relieve Earth of its most polluting industries. Americans would find their gilded paths on the high frontier owing to "the availability in the space habitats of high-paying jobs, of good living conditions, and of better opportunities" (O'Neill 263). This space-age disciple of Turner attracted keen attention for his visionary plan, including from *National Geographic*, which featured O'Neill's "High Frontier" in its 1976 bicentennial edition as a fitting corollary to the nation's pioneering history (Asimov 76–89).

Although Fletcher sent a copy of that issue of *National Geographic* to the White House, President Gerald Ford eschewed his entreaties and allocated just enough money so that a downsized NASA could develop the space shuttles as the platform for future spaceflight. So too did his successor

President Jimmy Carter, whose transition team warned against "an ersatz [space] frontier to replace the ones we have conquered and polluted" ("NASA Recommendations"). If Carter heeded that warning, his 1980 Republican rival Ronald Reagan did not as the president later announced that the shuttle and his proposed space station would help stimulate "Americans' pioneer spirit" and develop "our next frontier: space" (Boffey A1). That frontier motif was not yet White House and NASA vernacular in April 1981, when Space Shuttle Columbia commander and lunar-landing veteran John Young suggested that the shuttle's maiden "journey brought man a step closer to the stars." *Newsweek* magazine reported this was "not precisely what NASA's public-relations people wanted to hear," since they had "been promoting the shuttle in precisely the opposite way . . . a 'space truck' whose mission is not exploration but the exploitation of the familiar region of nearby space" (Adler et al. 28). However, Reagan's election and the arrival of America's shuttle fleet augured a revival of the frontier motif. This was clear when Reagan heralded the final "experimental" shuttle mission as "the historical equivalent to the driving of the golden spike which completed the transcontinental railroad" (Hannifin and Golden).

The space frontier motif became colloquial once again as NASA followed the White House's lead and depicted the shuttle as a frontier workhorse and its future space station as a pioneer outpost for workaday Americans. These spaceflight systems, the 1985 IMAX film *The Dream Is Alive* averred, meant that Americans "now know how to live and work in space." A half million students prepared to do so as members of the Young Astronauts Program, a math and science initiative endorsed by President Reagan, who praised its fledglings as the "generation that will move forward to harness the enormity of space" (Young Astronauts Council). This was the cohort invited by the new United States Space Camp, "as part of our first real space traveling generation," to come to camp and learn to "design spacecraft for interplanetary exploration, occupy space stations" and "become a vital link in the chain of people we need to take us beyond our own planet" (Schulke 13).

The shuttle astronauts who served as role models for this spacefaring generation were not the daredevils of yore "subjected to bone-crunching lift-offs or breathtaking splash-downs." *Time* explained they were a "new breed" whose ranks initially included men and women, Blacks, Jews, and a person of Asian descent (Hannifin et al.). This novel pedigree applied also to the brainy civilians who made up the bulk of mission and payload specialists, but even the many shuttle pilots who were military aviators did not fit the Mercury astronaut archetype. Unlike those heroic stick-and-rudder men, Tom Wolfe explained, shuttle pilots came of age when "automatic guidance systems [became] increasingly important" (Wolfe, "Columbia's Landing"). The shuttle's flight profile also meant that astronauts no longer needed the extreme conditioning and rugged desert and jungle training of their Mercury and lunar-landing forebears (Cooper 15). A children's book on Astronaut John Young illustrated this change, juxtaposing a picture of the youthful Apollo trainee wrapped in desert garb with a photo of the middle-aged shuttle commander peering through reading glasses. The book made clear that astronauts no longer needed the muscle or steely nerves Young exhibited on the Moon, for the shuttle came off as a routine cargo transporter that had taken the peril out of spaceflight (Westman).

As shuttle Astronaut Joe Allen plainly explained, the "astronauts of the inaugural era of space travel were explorers, pioneers who ventured briefly out to the fringe of a limitless frontier." Whereas they tended "to resemble cowboys, sitting in ejection seats instead of saddles, wearing helmets instead of wide-brimmed hats," Allen's colleagues were "their logical successors, men and women who go to space to work." The shuttle ended the "Spartan era of space travel" and allowed them to feel "increasingly at home in the new frontier" where they could don "sport shirts and slacks during their days in orbit" and "eat shrimp cocktail and barbecued beef and sleep in private bunks" (Allen 21, 31). NASA officials projected confidence in the shuttle's safety and prosaic accommodations by holding a national competition in 1984 for the "first private citizen passenger in the history of spaceflight." The winning candidate, New Hampshire high school teacher Christa McAuliffe, embraced the role of space-age homesteader and compared herself to "the pioneer travelers of the Conestoga wagon days" ("Remarks of the Vice President Announcing the Winner of the Teacher in Space Project"). NASA picked up this Turnerian thread in its lesson plans for schools tracking McAuliffe's flight, asking students "to compare our future space settlers and pioneers to the early settlers and pioneers of America" and consider whether "migrations from Earth to Space Stations and other planets will be similar to [their] migrations" (*Teacher in Space)*.

These frontier allusions reflected widespread optimism about the nation's spacefaring future, and they gained force from disconcerting indications that the glitch-prone space shuttle was not the "most reliable, flexible, and cost-effective launch system in the world" (*We Deliver)*. While NASA struggled to make the orbiter live up to that billing, the Soviet Union successfully operated an orbital station, and the European Space Agency deployed a more affordable launcher that became "the world's only successful commercial rocket system" (Canby 430; Osbourne 52, 57). These troubling affairs added to a general alarm that America's "once unchallenged preeminence in commerce, industry, science, and technological innovation is being overtaken by competitors throughout the world" (National Commission on Excellence in Education). Owing to the shuttle's unexpected failure to fulfill President Reagan's 1982 directive to "maintain United States leadership in space," he appointed a National Commission on Space (including Thomas Paine and Gerard O'Neill) "to formulate a bold agenda to carry America's civilian space enterprise into the twenty-first century" ("National Security Decision Directive Number 42"). Its 1986 report, *Pioneering the Space Frontier*, predictably called for the US to "lead the exploration and development of the space frontier, advancing science, technology, and enterprise, and building institutions and systems that make accessible vast new resources and support human settlements beyond Earth orbit." The report recapitulated Turner's economy-boosting frontier and proclaimed, "America can create new wealth on the space frontier to benefit the entire human community."

As *Pioneering the Space Frontier* went to print, the Space Shuttle Challenger undercut the report's rosy assumptions when it exploded during ascent, taking the lives of seven astronauts (including Christa McAuliffe). Americans discovered the terrifying reality that the shuttles could catastrophically fail. Post-Challenger surveys nevertheless indicated "a strong shift of public sentiment in favor of the space program generally and the shuttle program in particular," a bump in the polls that may have reflected a Rooseveltian impulse to honor the sacrifice of fallen pioneers by carrying on their noble endeavors (Miller). Declaring that "the future does not belong to the fainthearted," President Reagan assured a grieving nation that the brave "*Challenger* crew was pulling us into the future, and we'll continue to follow them" ("Address to the Nation on the Explosion of the Space Shuttle Challenger"). That bump may have also stemmed from an ongoing concern that "the US lead in space is being threatened as the Soviet Union continues its ambitious space program and Europe and Japan move aggressively to harvest the bounty of space" (*Space—America's New Competitive Frontier*). The Turnerian call found its most prominent spokesman in President George H. W. Bush, who encouraged Americans to look to outer space as "our nation's frontier, our manifest destiny." When announcing his Space

Exploration Initiative in July 1989, Bush hailed a familiar history from "the voyages of Columbus to the Oregon Trail to the journey to the Moon" to galvanize support for a "sustained program of manned exploration of the solar system and, yes, the permanent settlement of space" ("Remarks on the 20th Anniversary of the *Apollo 11* Moon Landing").

President Bush summoned the public enthusiasm evident in polls, but that devotion to the space frontier faded fast, as did public faith in the shuttle and American technological wizardry. NASA admitted that the Challenger accident "brought to an end over three decades of success of what was perceived to be an almost superhuman ability on the part of the American production machine to set a goal, meet it and then with equal resolve exceed that goal and set another" (*Regaining the Competitive Edge*). The disaster also put the shuttle program on hold for two years, after which US military priority over shuttle missions lowered its public profile. That profile dropped further when the Cold War dramatically ended and Japan's economy faltered, leaving the US largely unchallenged on the world stage in the 1990s and thereby ending the national challenges that had fueled a renewed space race and sustained the space frontier motif. Then in the aftermath of another shuttle disaster—Columbia's seven crew members perished as the shuttle burned up on reentry in 2003—NASA began phasing out the shuttle program and from 2014 until 2020 relied on Russian partners to deliver astronauts to the orbiting International Space Station (ISS). In November 2020, SpaceX became the first among a growing number of commercial spaceflight companies with interests in space tourism to ferry astronauts to the ISS. This suggests their industry, rather than the federal government, will kick-start a new space age and open the high frontier once and for all.

Pioneering Privateers

In what may be the first study of "extraterrestrial tourism," aerospace engineer Krafft Ehricke predicted in 1967 that it might "evolve in the wake of" a national undertaking that "won geospace 'territory' for the benefit of all" (Ehricke). In other words, tourism might follow only after the conquest of the space frontier. Fifty years later, an aerospace industry blogger reversed Ehricke's chronology by crediting SpaceX's plans "to create space tourism" for "pushing the boundaries of the final frontier." In this calculus, the company's aim to send tourists to the ISS for as much as $50 million and around the Moon for many times more would enable it to build "a sustainable mode of transportation from Earth to the [M]oon and Mars . . . and eventually lead colonization efforts" in space (SPACECOM website). This private transportation system would position SpaceX to thrive in a $300 billion annual market for commercial space services (Weinzierl, Lambright). Although most of that market is for satellite launches, the potential $3 billion space tourism sector has attracted outsized public attention due to the drama of human spaceflight and the likelihood that tourists may soon outnumber NASA astronauts.

Public attention has also followed pronouncements from charismatic businessmen like SpaceX CEO Elon Musk, who floated remarkable timelines for private travel to Earth orbit (initially projected as 2020), the Moon (2022), and Mars (2024). Musk may have done so to boost stock value by pitching his company's readiness to deploy effective launch systems and achieve its visionary (and profitable) mission "to revolutionize space technology, with the ultimate goal of enabling people to live on other planets" (SpaceX website). Blue Origin founder Jeff Bezos may have done the same, and burnished his Amazon-founding reputation for game-changing entrepreneurship, by announcing plans to land people on the Moon by 2024 before turning toward the grander project of building space colonies like those conceived by Gerard O'Neill (Powell, Blue Origin website). Although Richard Branson has a lower-altitude target, he declared "the dawn of a new Space Age" when announcing the recent IPO for his suborbital space tourism company, Virgin Galactic (Jolly). Like his billionaire peers, Branson may be impelled not only by profits but also by his company's lofty vision of developing hardware—absent NASA doing so—that will "catalyze a new age of space exploration which promises enormous positive potential" for humankind (Virgin Galactic website).

Many spaceflight advocates use the trope of the frontier as they, too, elevate the nascent space tourism industry as a progressive force for humanity (V. Smith, Globus). Apollo Astronaut Buzz Aldrin, who has long promoted the space frontier as critical to human welfare, described "commercial space tourism" (again, absent public funding) as a necessary catalyst for technologies essential for "longer-range civil programs such as a return to the Moon and a sustained program of Mars exploration and settlement" (Aldrin and Jones 178). A veteran shuttle astronaut, who similarly believes the "survival of the human race" is at stake, complimented the likes of Musk, Bezos, and Branson for building the infrastructure that will make "the dream of 'space travel for the masses' a reality," characterizing that infrastructure as "a necessary step in the eventual expansion of humanity into space" (Pelton and Marshall 8, 11). US Senator Ted Cruz (R-Tex) did so as well in his 2017 hearing, "Reopening the American Frontier," when he praised these entrepreneurs for being "crazy enough to think they can change the world" by "sparking a renaissance in space" through private investment and innovation ("Senator Cruz: 'Our National Space Program is on the Verge of a Renaissance'").

These proponents of a third phase of human spaceflight, led by private industry (albeit supported by government subsidies) and leveraged by market demand for high-end travel, contend that an incipient space tourism industry is ready to break out—this nearly twenty years after Space Adventures Ltd. brokered the first paid berth on a Russian Soyuz to the ISS. Aside from the irony that erstwhile communist Russians, rather than American capitalists, rocketed seven tourists to the ISS between 2001 and 2009 for nearly $20 million each, these flights indicated that many of the world's richest people were ready to spend huge sums on such a novel experience. Now that US companies like SpaceX are poised to launch people into orbit and NASA will accept fee-paying tourists at the ISS, Space Adventures says it is ready "to open the space frontier to private citizens" who want to "circle the Earth, or travel beyond the Earth" (Space Adventures website). The merely wealthy who cannot afford this frontier adventure are lining up to fly for just $250,000 on Virgin Galactic's *SpaceShip Two*, hoping to follow owner Richard Branson's July 2021 test flight to the edge of space. With a recent market value of nearly $1 billion and a taxpayer-financed spaceport ($250 million) in New Mexico, Virgin Galactic seeks to control a potential market of more than a million customers and become the "Spaceline for Earth."

While scholars speculate about the industry's viability (Launius and Jenkins), a *Wall Street Journal* reporter was quite sure that "Shooting Rich People into Space Isn't the Final Frontier" (Sindreu). This provocation speaks to the dissonance of the space frontier motif when applied to high-end commercial space travel. Industry boosters do so to excite public interest in their commercial endeavors, and they may sincerely believe their businesses operate to open frontiers for humankind as well as earn profits. This is especially true of deep-pocket visionaries like Musk and Bezos, whose start-ups have disrupted commerce on Earth and who apparently seek lasting legacies by disrupting state-managed spaceflight as well. But the motif emerged during a more liberal era of state-managed spaceflight, when NASA coordinated a collective national project of hundreds of thousands of workers and millions of supportive citizens to putatively expand freedom, democracy, and prosperity. This narrative conformed to the mythology of American frontier expansion, which focused on explorers and yeoman settlers rather than moneyed interests who built the railroads or toured the West as thrill-seeking tourists. Thus, shooting rich people into space does not resonate with the egalitarian cant of this nationalist storyline of everyday pioneers working to sustain democracy. Nor does it align with the frontier trope's promise of widely shared liberty and prosperity. Although Americans have increasingly idolized the pursuit of wealth as the US adopted a more neo-liberal economy—in which even essential public goods are provided by private enterprise (including space launch services)—their nationalist mythology has not evolved to reckon with this political economy and unprecedented inequality. It certainly has not morphed to celebrate the concentration of wealth that enables only a select few to purchase exorbitantly priced tickets to space. On the contrary, since inequality now inflames liberals on the left and populists on the right, rich space tourists are more likely to remind workaday Americans of their narrowed economic opportunity and freedom than to come off as heroic exemplars of a pioneering nation. Witness the wide-ranging derision of billionaire Jeff Bezos, whose July 2021 suborbital flight on his Blue Origin spaceship contrasted with the economic fallout experienced by millions of Americans during an ongoing global pandemic.

Finally, the rhetorical challenge faced by space tourism boosters goes deeper still. In searching for a narrative that can excite the public—and its support for state and federal subsidies needed to sustain their business model—they often rely on a timeworn trope developed during the first phase of human spaceflight in the 1960s and fractionally revived during the second phase in the 1980s. Since then the space frontier motif has lost further saliency as American culture has changed and the national conversation has fractured among multiplying media venues, undercutting what once appeared to be public consensus about the nation's character and purpose. This was already evident in the early 1990s when the *Washington Post*, usually a cheerleader for NASA, skeptically discounted the space frontier motif and questioned the nationalist narrative "that we're pioneers, that we explore frontiers, that we use technology in that pursuit, and that we are a country with a special sense of our place in history" (Carlson 22).

It was even more apparent when a space evangelist took to the pages of *USA Today* in 2010 to express hope for a new space program that could inspire Americans "to reimagine a frontier that will allow us to open up this hunkered-down existence." He could have invested that hope in an incipient space tourism industry but did not do so because Americans were stuck "between mythologies," unable to see the path forward in space because "we are not what we once were . . . and we do not yet know what we will become" (Deutsch 70).

Proponents of commercial spaceflight are thus also stuck between mythologies. They are compelled to channel inherited discourses and market that industry as the new pioneering vanguard of a space frontier. But that motif has never comfortably included commercial actors, and it may no longer move American audiences.

Works Cited

"Address to the Nation on the Explosion of the Space Shuttle Challenger," *Public Papers of the Presidents of the United States.* Administration of Ronald Reagan, 1986.

Adler, Jerry, et al. "In Space to Stay." *Newsweek*, 27 Apr. 1981, p. 28.

Aldrin, Buzz, and Ron Jones. "Changing the Space Paradigm: Space Tourism and the Future of Space Travel." *Space: The Free-Market Frontier*, edited by Edward Hudgins, Cato Institute, 2002, p. 178.

Allen, Joseph. *Entering Space: An Astronaut's Odyssey*. Stewart, Tabori, & Chang, 1984.

"Angel of Mercy to the Astronauts." *Ebony*, June 1966, pp. 49–52.

"Annual Message to the Congress on the State of the Union," 20 January 1972. *Public Papers of the Presidents of the United States.* Richard M. Nixon, 1972, p. 71.

Asimov, Isaac. "The Next Frontier?" *National Geographic*, July 1976, pp. 76–89.

Berkner, Lloyd V. House Subcommittee on Territorial and Insular Affairs, 10 June 1960. Library of Congress Manuscript Division, Lloyd V. Berkner Papers, Box 21, Statement on Antarctica.

Blue Origin. https://www.blueorigin.com/fly-with-us/.

Boffey, Philip M. "President Backs U.S. Space Station as Next Key Goal." *New York Times*, 26 Jan. 1984, p. A1.

Brown, Robert McAfee. "Moon Shot and Afterthoughts," *Christianity and Crisis*, vol. XXIX, no. 15, 15 Sept. 1969.

Bush, Vannevar. Remarks at Convocation of MIT, "The Essence of Security," December 5, 1949, Library of Congress Archives, *Papers of Vannevar Bush*, Box 132, file Speech, MIT 12/5/49.

———. *Science: The Endless Frontier*, US Government Printing Office, 1945, Foreword.

Canby, Thomas. "Are the Soviets Ahead in Space." *National Geographic*, October 1986, pp. 52, 57.

Carlson, Peter. "Is NASA Necessary?" *Washington Post Magazine*, 30 May 1993, p. 22.

Carpenter, Scott, et al. *We Seven*. Simon and Schuster, 1962.

Chang, Kenneth. "SpaceX Plans to Send 2 Tourists Around the Moon in 2018." *New York Times*, 27 Feb. 2017.

Cooper, Henry S.F. *Before Lift-Off: The Making of a Space Shuttle Crew*. Johns Hopkins UP, 1987.

Deutsch, Bob. "Saving Our Space Program," *USA Today Magazine*, November 2010. Reprinted in *American Space Exploration and Development*, edited by Christopher Mari, Wilson, 2011, p. 70.

Ehricke, Krafft. "Space Tourism," American Astronautical Society, Commercial Utilization of Space, 13th Annual Meeting, May 1967. Smithsonian National Air & Space Museum Library, OS-171400–08, Documents Social Impact, Tourism, Reports.

Fisher, Allan. "Exploring Tomorrow with the Space Agency." *National Geographic*, July 1960, pp. 52, 59.

Glenn, John, Jr. *"P.S. I listened to your heartbeat": Letters to John Glenn*. World Book Encyclopedia Science Service, 1964.

Glennan, T. Keith. Speech to the US Air Force. 24 August 1959. National Archives, RG 359, Executive Office of the President Office of Science and Technology, Subject Files 1957–62, Box 65, file "Space-Nat'l Aeronautics & Space Admin."

Globus, Al. "A radically easier path to space settlements." *The Space Review*, 27 Feb. 2017.

Graham, Billy. Letter to President Dwight D. Eisenhower, 2 December 1957. NASA History Office, file "Impact, Public Opinion: Sputnik," 006737.

Hannifin, Jerry, and Frederic Golden. "Space: The Once and Future Shuttle." *Time*, 19 July 1982.

Hannifin, Jerry, et al. "Space: Sally Joy Ride into the Sky." *Time*, 13 June 1983.

Horrigan, Brian. "Popular Culture and Visions of the Future in Space, 1901–2001." *New Perspectives on Technology and American Culture*, edited by Bruce Sinclair, American Philosophical Society, 1986, pp. 49–67.

"Investment in the Future." *New York Times*, 8 Jan. 1972.

Jolly, Jasper. "Virgin Galactic Launches on the New York Stock Exchange." *The Guardian*, 28 Oct. 2019, https://www.theguardian.com/science/2019/oct/28/virgin-galactic-spce-launches-new-york-stock-exchange-richard-branson.

Lambright, W. Henry. "Launching commercial space: NASA, cargo, and policy innovation." *Space Policy*, 2015, pp. 23–31.

Launius, Roger, and Dennis Jenkins, "Is It Really Time for Space Tourism?" *Astropolitics* 4: 2006.

"Man in Space," March 9, 1955. *Walt Disney Treasures: Tomorrowland*, Walt Disney Video, 2004.

McCurdy, Howard E. *Space and the American Imagination*. Smithsonian Press, 1997.

"Men of the Year." *Time*, 3 Jan. 1969, p. 4.

Merchant, Livingston. *Review of the Space Program*, House Committee on Science and Astronautics. 86th cong. 2nd sess., 20, 22, 25–29 January, 1–5 February 1960, p. 3.

Michael, Donald N. "The Beginning of the Space Age and American Public Opinion." *Public Opinion Quarterly*, 1960, pp. 573–582.

Miller, Jon D. *The Impact of the Challenger Accident on Public Attitudes Toward the Space Program: A Report to the National Science Foundation*. 25 Jan. 1987, p. i.

NASA. *Space: The New Frontier*, EP-6, 1966.

NASA and the Now Syndrome. NASA, 1975, p. 2.

"NASA Recommendations." Nick MacNeil memo, Carter-Mondale Transition Planning Group, 15 Jan. 1977. Jimmy Carter Library, Stern Files, Box 4, file "NASA."

National Commission on Excellence in Education. *A Nation at Risk: The Imperative for Educational Reform*. US Government Printing Office, 1983.

"National Security Decision Directive Number 42, 'National Space Policy,' 4 July 1982." *Exploring the Unknown: Selected Documents in the History of the U.S. Civil Space Program*, edited by John Logsdon et al., NASA, 1995, p. 591.

Nauton, Joan Ena. "A Woman in Outer Space? Good Idea–But Still a Joke." *Miami Herald*, 21 May 1972, p. 17.

Niebuhr, Reinhold. "After Sputnik and Explorer." *Christianity and Crisis* vol. XVIII, no. 4, 17 March 1958, p. 30.

"On Courage in the Lunar Age." *Time*, 25 July 1969, p. 19.

O'Neill, Gerard. *The High Frontier: Human Colonies in Space*. Morrow, 1977.

Osbourne, David. "Business in Space." *Atlantic Monthly*, May 1985.

Pelton, Joseph, and Peter Marshall. *License to Orbit: The Future of Commercial Space Travel*. Apogee Books, 2009.

Pioneering the Space Frontier: The Report of the National Commission on Space. Bantam Books, 1986.

Powell, Corey. "Jeff Bezos foresees a trillion people living in millions of space colonies." *NBCNews.com*, May 15, 2019, https://www.nbcnews.com/mach/science/jeff-bezos-foresees-trillion-people-living-millions-space-colonies-here-ncna1006036.

Regaining the Competitive Edge. NASA, 1987.

"Remarks by James Fletcher to the Salt Lake Rotary Club," 6 June 1972. National Air and Space Museum Archives, file OS-170889–56.

"Remarks of the Vice President Announcing the Winner of the Teacher in Space Project," 19 July 1985, *Public Papers of the Presidents*, Administration of Ronald Reagan, 1985, p. 932.

"Remarks on the 20th Anniversary of the *Apollo 11* Moon Landing," 20 July 1989, *Public Papers of the Presidents*, Administration of George Bush, 1989, p. 992.

Roosevelt, Theodore. "The Strenuous Life," Speech before the Hamilton Club, Chicago, 10 April 1899.

Rudder, James Earl. "Preparation of the Whole Man." Nuclear Space Seminar, Amarillo, Texas, August 1962. Library of Congress Archive, Lloyd V. Berkner Papers, Box 37, file "Space Technology and You."

Schulke, Flip. *Your Future in Space: The U.S. Space Camp Training Program*. Knopf Books for Young Readers, 1986.

Scott, Stan. Memo to Ken Clawson, "NASA's Image," 19 February 1974. National Archives, Nixon Presidential Material Project, White House Central Files, FG 164 NASA, Box 2, file "[EX] Fg 164 NASA."

"Senator Cruz: 'Our National Space Program is on the Verge of a Renaissance.'" *States News Service*, 13 July 2017.

Sindreu, Jon. "Shooting Rich People Into Space Isn't the Final Frontier." *Wall Street Journal*, 10 July 2019.

Smith, Michael. "Selling the Moon: The U.S. Manned Space Program and the Triumph of Commodity Scientism." *The Culture of Consumption: Critical Essays in American History, 1880–1980*, edited by Jackson Lears et al., Pantheon Books, 1983, p. 200.

Smith, Valene. "Space Tourism: The 21st Century 'Frontier.'" *Tourism Recreation Research*, v. 25(3), 2000, pp. 5–15.

———. "Soviet Claiming Lead in Science." *New York Times*, 5 Oct. 1957, p. 2.

———. *Soviet Man in Space*. Moscow: Foreign Languages Publishing House, 1961, p. 12.

Space Adventures website, https://spaceadventures.com/about-us/our-vision/.

Space: America's New Competitive Frontier. Business-Higher Education Forum, April 1986, p. ii.

"Space Doctor for the Astronauts." *Ebony*, April 1962, pp. 35–39.

SpaceX website, https://www.spacex.com/about.

Spiller, James. *Frontiers for the American Century: Outer Space, Antarctica, and Cold War Nationalism*. Palgrave, 2015.

"Statement about the Future of the United States Space Program," 7 March 1970. *Public Papers of the Presidents:* Richard Nixon, 1970, pp. 250–251.

"Statement by George Mueller," 24 July 1969. Gerald R. Ford Library, H. Guyford Stever Papers, 1930–1990, Box 66, file "NASA, 1967–70.

"Statement by President Nixon," 5 January 1972, *Public Papers of the President*, Richard M. Nixon, 1972.

"Statement by Vice President Spiro Agnew," Chairman of the National Aeronautics and Space Council, 21 May 1969. NASM, OS-170884–50.

Teacher in Space. NASA, 1985.

The Dream Is Alive. IMAX, Lockheed, and Smithsonian Institution, 1985.

Turner, Frederick Jackson. "The Significance of the Frontier in American History" (1893). *Rereading Frederick Jackson Turner*, edited by John Mack Faragher, Yale UP, 1994, pp. 31–32.

Van Dyke, Gretchen J. "Sputnik: A Political Symbol and Tool in 1960 Campaign Politics." *Reconsidering Sputnik: Forty Years Since the Soviet Satellite*, edited by Roger Launius et al., Harwood Academic, 2000, p. 396.

Virgin Galactic website, https://www.virgingalactic.com/vision/.

Von Braun, Wernher. "Why Space Exploration is Vital to Man's Future." *Space World*, September 1969, pp. 31–33.

We Deliver, NASA, 1983, p. 2.

Website of SPACECOM America's Space Conference and Exposition, https://spacecomexpo.com/spacex-pushing-boundaries-final-frontier/.

Weinzierl, Matthew. "Space, the Final Economic Frontier." *Journal of Economic Perspectives* vol. 32, no. 2, Spring 2018, pp. 173–192.

Weitekamp, Margaret. *Right Stuff, Wrong Sex: America's First Women in Space Program*. Johns Hopkins UP, 2004.

Westman, Paul. *John Young: Space Shuttle Commander*. Dillon Press, 1981.
Wolfe, Tom. *The Right Stuff*. Bantam Books, 1979, p. 71.
———. "Columbia's Landing closes a Circle." *National Geographic*, October 1981, p. 477.
"Woman Space Pioneer." *Philadelphia Inquirer*, 15 Jan. 1959.
Young Astronauts Council, *Young Astronaut Program Fact Sheet*, 1988; Ronald Reagan letter to Young Astronauts on the Foundation's first anniversary, October 17, 1985.

Contributors

J BRET BENNINGTON is professor of geology and department chair in the Department of Geology, Environment, and Sustainability at Hofstra University, where he has taught for twenty-nine years. His research and teaching interests include paleontology, the geology of Long Island, and Charles Darwin and the history of geology. In 2019 Dr. Bennington was presented with the Neil Miner Award for "exceptional contributions to the stimulation of interest in the Earth sciences" by the National Association of Geoscience Teachers. In addition to his research publications in paleontology, sedimentology, and earth science education, he has edited two other conference volumes, *Learning from the Impacts of Superstorm Sandy* (coeditor with E. Christa Farmer) and *The 1930s: The Reality and the Promise* (coeditor with Zenia DaSilva and Michael D'Innocenzo).

MATTHEW H. HERSCH is associate professor of the history of science at Harvard University. He received a JD from New York University School of Law and a PhD in the history and sociology of science from the University of Pennsylvania, where he taught in the School of Arts and Science and the School of Engineering and Applied Science. He has held fellowships at the Smithsonian Institution, NASA, the Huntington-USC Institute on California and the West, the Heyman Center for the Humanities at Columbia University, and the Institute for Advanced Study. He is the author of *Inventing the American Astronaut*, coauthor (with Ruth Schwartz Cowan) of *A Social History of American Technology*, 2nd Edition, and coeditor (with Cassandra Steer) of *War and Peace in Outer Space: Ethical and Legal Boundaries*.

RODNEY F. HILL is associate professor and department chair in the Department of Radio, Television, Film at Hofstra University's Lawrence Herbert School of Communication. He is co-author of *The Encyclopedia of Stanley Kubrick* and a contributor to several other books, including *The Essential Science-Fiction*

Television Reader, The Stanley Kubrick Archives, After Kubrick: a Filmmaker's Legacy, and *The Bloomsbury Companion to Stanley Kubrick*.

ROSANNA PEROTTI is professor of political science at Hofstra University, where she teaches classes in US political institutions, public opinion, elections, and immigration policy. She has edited or co-edited seven books stemming from Hofstra University's presidential conferences and has authored articles and conference papers on US immigration policy, student political participation and civic engagement, and pedagogy.

CHRIS ROBINSON is assistant professor of film and media studies at the University of Arkansas Little Rock and resident curator and programmer for the Arkansas Cinema Society. He is also a curator for the Telluride Film Festival and a consultant for Turner Classic Movies. He has taught at Emerson College, the University of Kansas, and Columbia University. He has worked in production, distribution, and exhibition, and also has served as post-production editor on numerous documentaries for HBO, ESPN, and PBS, including films that screened at the Sundance Film Festival.

PATRICIA ROSSI is an attorney and an independent scholar. She has presented numerous academic papers over the years, at university conferences within the New York tri-state region. Her academic articles, narrative essays, and poetry have been published in a variety of magazines, newspapers, and journals. Her academic essay analyzing the socio-economic impact of the fashion industry was included in *The 1930s: The Reality and the Promise*. She serves on a variety of not-for-profit boards as well as on her college alumni board. She also volunteers as a "writing to heal" facilitator for cancer patients at a local hospital and cancer coalition.

SABRINA G. SOBEL earned a BA in chemistry from Pomona College (1987) and was awarded the Rowan Memorial Award for being "the student who shows the best promise of making the most contributions to Chemistry." She earned her PhD in inorganic chemistry from the University of California, Berkeley (1993), where she completed her thesis work partially under Dr. W. Armstrong, and as a guest researcher in Dr. E. Stiefel's group at IBM Almaden Research Center, New Jersey. Dr. Sobel has spent her career at Hofstra University as the sole inorganic chemist in the department. She has mentored both undergraduates and high school students in research and served as the chair of the Department of Chemistry at Hofstra from 2007 to 2016. Her research is an eclectic mix of the study of biologically relevant zinc compounds, oscillating chemical reactions, aluminum corrosion, and chemical education. Dr. Sobel has served on three rounds of development of ACS standard undergraduate inorganic chemistry exams. In addition to her teaching duties, she is currently director of General Chemistry Lab classes at Hofstra.

JAMES A. SPILLER is professor of history at SUNY Brockport, the State University of New York, where he teaches modern US History with specialties in American culture and politics, science and technology, and environment. His recent work includes *Frontiers for the American Century: Outer Space, Antarctica, and Cold War Nationalism*, "International Science in Antarctica," in Hugh Slotten et al., eds., and *Modern Science in National, Transnational, and Global Context.*

ALLISON WHITNEY is associate professor of film and media studies in the Department of English at Texas Tech University. Her research focuses on discourses of film and technology, sound studies, film and media studies pedagogy, and film genres, including science fiction. Publications include "Formatting Nostalgia: IMAX Expansions of the Star Wars Franchise" in *Star Wars and the History of Transmedia Storytelling* and "Cinephilia Writ Large: IMAX in Christopher Nolan's *The Dark Knight* and *The Dark Knight Rises*" in *The Cinema of Christopher Nolan.*

JULIE WOSK is professor emerita, State University of New York, Maritime College. She is the author of several books about the social impact of technology, including *Women and the Machine: Representations From the Spinning Wheel to the Electronic Age*, which has a chapter on women and aviation; and *My Fair Ladies: Female Robots, Androids, and Other Artificial Eves*. She is also an independent museum curator whose traveling exhibit "Imaging Women in the Space Age" was first shown at the New York Hall of Science in Queens.

Index

Page numbers in *italics* refer to illustrations.